IMAGES
of America

CAMP UPTON

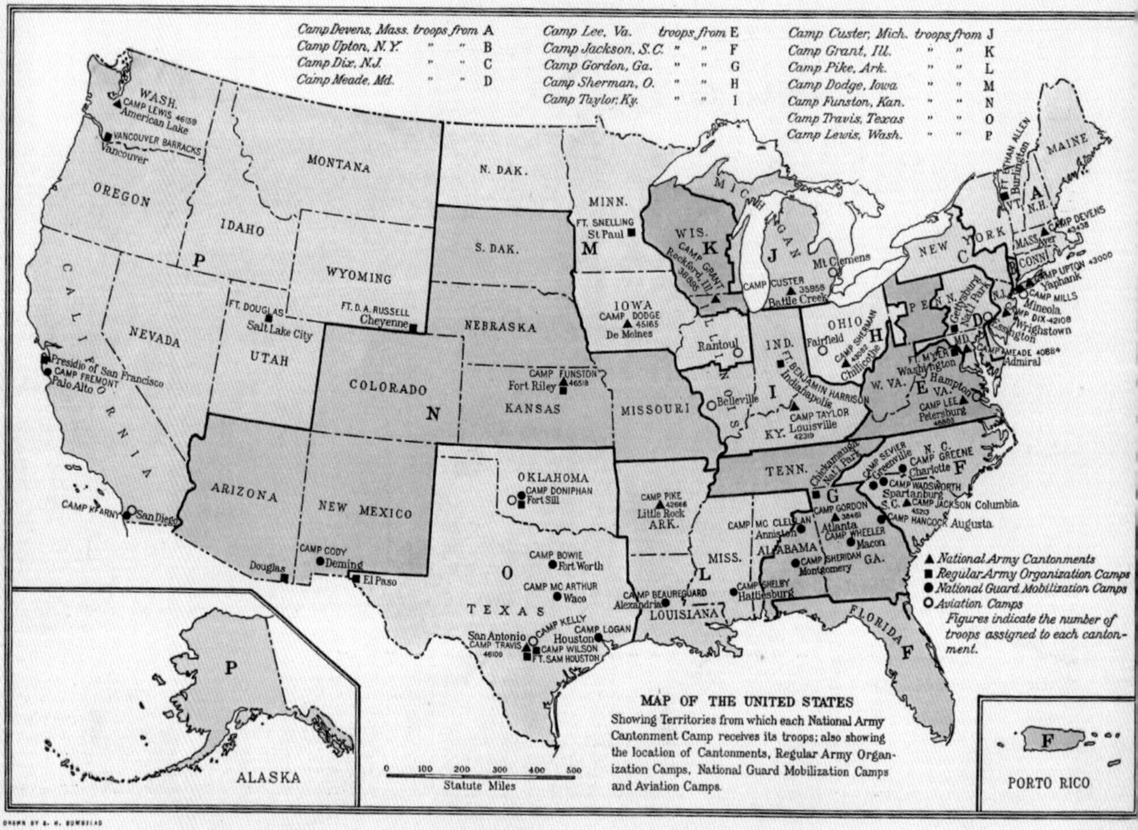

This map, from the November–December 1917 issue of *National Geographic*, shows the locations of the 16 National Army cantonments across the country that were built in 1917. Most were located along the East Coast to facilitate deployment of troop ships to Europe. Camp Upton was first known as Camp Long Island and Camp Yaphank. (Courtesy of the Longwood Public Library, Thomas R. Bayles Collection.)

On the Cover: The 306th Field Artillery regiment was part of the 77th Division (the "Liberty" Division), formed primarily of soldiers from the New York metropolitan area who trained at Camp Upton on Long Island during World War I. This card was mailed on December 5, 1917, to Mrs. Frank. J. Steuterman in Memphis, Tennessee, from Adolph, 306th Field Artillery, Company A, the "first man kneeling on the right hand side." (Courtesy of Paul Infranco.)

IMAGES
of America

CAMP UPTON

Suzanne Johnson and David Clemens

ARCADIA
PUBLISHING

ISBN 978-1-4671-2753-0

Published by Arcadia Publishing
Charleston, South Carolina

Printed in the United States of America

Library of Congress Control Number: 2017941299

For all general information, please contact Arcadia Publishing:
Telephone 843-853-2070
Fax 843-853-0044
E-mail sales@arcadiapublishing.com
For customer service and orders:
Toll-Free 1-888-313-2665

Visit us on the Internet at www.arcadiapublishing.com

We dedicate this volume to Henry James Jones of Yaphank, who served in Company B, 305th Machine Gun Battalion, and was killed in the Argonne Forest on November 1, 1918.

Contents

Acknowledgments

The authors are grateful to the Yaphank Historical Society and Pres. Robert Kessler for agreeing to sponsor this publication, which developed due to interest in the 100th anniversary of the United States' entry into World War I. Specifically, interest in what had been Camp Upton, located just a few miles from the Longwood Public Library, provided the initiative to explore the extensive collection of images, diaries, regimental histories, and newspapers available in the library's Thomas R. Bayles Local History Collection. The library values its role in preserving local history, and with the support of the Board of Trustees, it has continued to enhance the collection over the past 30 years. Thomas Bayles, a lifelong Middle Island resident, published a brief history of Camp Upton in 1974 and revised it in 1977. His writing has been a source of inspiration for this project as well as the outstanding website *Longwood's Journey*, created by Longwood Middle School social studies teachers Paul Infranco and Victor Massian and their students. The website makes available digitized copies of photographs, letters, diaries, *Trench and Camp* and *Upton Nooz* (the Camp Upton newspapers), and many other primary source materials. Paul Infranco's work with Tom Bayles's son, Donald, has broadened our understanding of the camp's history. Don Bayles's uncle, Albert Bayles, worked at the camp during construction, and Tom worked in the freight office of the Long Island Rail Road there. In 1923, Albert and Tom's father, Richard Bayles, produced a record from a scrapbook of those serving in World War I from Brookhaven Town. Such firsthand knowledge of the history of Camp Upton has been invaluable.

Unless otherwise noted, all images appear courtesy of the Thomas R. Bayles Collection at the Longwood Public Library. Additional photographs were lent by Paul Infranco and obtained from the Library of Congress. The archives of the Yaphank Historical Society provided key primary source material. Mark Rothenberg's exhaustive Long Island history website at the Patchogue Medford Library, Dr. Tim Green at Brookhaven National Laboratory, and Brookhaven town historian Barbara Russell were of great help. Digitized photographs and documents on the library website longwoodlibrary.org, maintained by local history librarians Cynthia Sciacca and Melanie Cardone-Leathers, continue to add to our knowledge of the camp's evolution from 1917 through World War II to its current occupant, Brookhaven National Laboratory. Finally, the Longwood community has dedicated itself to honoring its veterans of all wars and conflicts in order to educate the next generations so that history need not repeat itself. This publication was created to further that goal. All authors' royalties produced by this book will go to the Yaphank Historical Society.

INTRODUCTION

It was called "the great war to end all wars." The Great War, later dubbed World War I, developed following a series of tragic, troubling incidents in Europe. The British luxury liner *Lusitania* was torpedoed by a German submarine in May 1915 off the coast of Ireland, with 128 Americans among the 1,198 who died. Pres. Woodrow Wilson, whose second-term slogan in 1916 was "he kept us out of war," soon found himself faced with a difficult decision. Wilson declared war on Germany on April 6, 1917, with the pledge that it would "make the world safe for democracy." The US Army was ranked 17th in the world and had just 25,000 soldiers. A new National Army needed to be raised and trained quickly. Wilson signed the Selective Service Act, and nine and a half million men registered for the draft. Two million would ultimately be trained to fight. Wilson selected Gen. John Pershing to command the American Expeditionary Forces (AEF) and go to Europe to assist the allied French and British armies, which had been fighting the Central Powers on the Western Front since 1914.

Not everyone was in support of going to war. At home, American women suffragists who had been fighting to win the right to vote marched in opposition to the war. Many feared the war effort would derail the suffrage movement. Eventually, Wilson supported the women's right to vote amendment, and suffragists in turn helped Wilson's war campaign. The Committee on Public Information campaign documented activities as the "doughboys" were being trained and supplied hundreds of images of Camp Upton and other cantonments. The nickname "doughboys" may have come from Army activity in Texas, where the white dust of adobe soil settled on US Infantry troops along the Rio Grande as they chased Pancho Villa along the Mexican border. They were called "adobes," shortened to "dobies," and then "doughboys," according to Laurence Stallings's 1963 history of the war.

The new and expanded army was made up of three types of troops, and they were prepared for overseas service in three different ways. Regular Army units were assigned to organizational camps and were formed into Divisions 1 to 8. National Guard units were assembled in mobilization camps and formed Divisions 26 to 42. Draftees were assigned to one of 16 regional cantonments named for Army generals to be inducted, trained, and organized into Divisions 76 to 92. National Guard units from 26 states were assembled into the 42nd Division, also known as the Rainbow Division, at Camp Mills in Mineola on Long Island. This division was sent overseas in November 1917 and received battlefield training in France. Draftees from the New York metropolitan area were assigned to Upton Cantonment, which became known as Camp Upton. US Army general J. Franklin Bell was assigned to Camp Upton with orders to create and train the new 77th Division. The vast majority of the units that were trained at Camp Upton were assigned to this division. The division was composed of four infantry regiments, the 305th, 306th, 307th, and 308th; three field artillery regiments, the 304th, 305th, and 306th; three machine gun battalions, the 304th, 305th, and 306th; the 302nd Engineers; the 302nd Field Signal Battalion; the 302nd Supply Train; the 302nd Ammunition Train; the 302nd Sanitary Train; and military police, a total of about 25,000 men and officers.

The site of the camp was carefully chosen. The requirements for a suitable site for the camp included proximity to railroad transportation and to the departure points around New York City, a warm climate, and large, undeveloped acreage. Hal B. Fullerton, special agent of the Long Island Rail Road, suggested the site to the Army, noting the access to three railroad stations: Wading River, Yaphank, and Manorville. After studying US Geological Survey maps, representatives from the Army and the Long Island Rail Road met out in the wilderness two miles from the quaint village of Yaphank and struck a deal. The largest landowners of property that became the camp included Young & Metzner Realty Company with 8,400 acres, Archibald Weeks with 3,200 acres, and North Shore Development with 2,400 acres. Years later, when the camp closed and the government purchased the property from these owners, the cost was just over $550,000. However, the cost of constructing the camp exceeded $6 million.

Not everything worked out according to the planned design. The orientation of the camp was changed so that the winds from the stables on the east would not blow over the barracks. However, that shift caused problems with drainage that needed to be resolved. Mosquitoes from the swampy areas to the south plagued the early days at the camp. Dr. William Gorgas, surgeon general of the US Army, advocated for the draining of ponds and swamps, fumigation, mosquito netting, and public water systems to remediate the problem.

After adding 5,000 acres for a rifle range, the camp boundaries extended from the Long Island Rail Road branch in Wading River in the north to South Country Road in the south and from the Peconic River in the east to the Carmans River in the west. The camp would ultimately cover 19,900 acres. It was described as a "soldier city," serving its important purpose for a short time and then quickly reverting to forest until the site was needed as a training camp once again 20 years later. In May 1919, the cemetery at Camp Upton was decertified, and the graves were moved from there and nearby Yaphank Cemetery to Cypress Hills Cemetery in Queens. The Army left behind 32 miles of roads, water mains, sewage facilities, and electric lines that were utilized for the next 90 years.

Following the auction and dismantling of the camp in 1921, the government unsuccessfully tried to sell the property. From 1923 to 1933, the property was used by the 198th Coast Artillery from Delaware for antiaircraft surveillance. Over 2,000 acres of the former camp was designated Upton National Forest by Pres. Calvin Coolidge in 1925. Coolidge at the time believed the forest would eventually serve as New York City's chief timber resource.

The camp property went through a series of other uses. In 1931, the site was considered for a state prison or a state mental hospital at the suggestion of Gov. Franklin D. Roosevelt. In 1932, the War Department offered the camp for sale but received only one inadequate bid. In 1934, the Army tried unsuccessfully to transfer it to the state. A use was finally found when a Civilian Conservation Corps (CCC) camp was established in the summer of 1933. Three camps employing 800 men worked at the CCC camp in tents. One project employed 18 men from Yaphank to dig up several hundred white pines and white birches and transplant them at Mitchell Field as part of a Civil Works Administration project during the Depression. The CCC camp closed in 1936, and Camp Upton was again abandoned. But when war approached, the camp was rebuilt in October 1940 to accommodate 20,000 soldiers and began accepting conscripts under the supervision of Col. C.W. Baird. In 1944, a convalescent hospital and rehabilitation center for wounded veterans was opened, and it operated for two years. For a short period, an alien detention camp existed there. A prisoner of war camp held 500 German prisoners in 1945. Finally, in 1946, the Army no longer had use for the camp and declared it surplus property. In 1947, Brookhaven National Laboratory was established as a research center for the peaceful use of nuclear energy on 5,000 of the camp's original 19,900 acres.

One

Out of the Wilderness

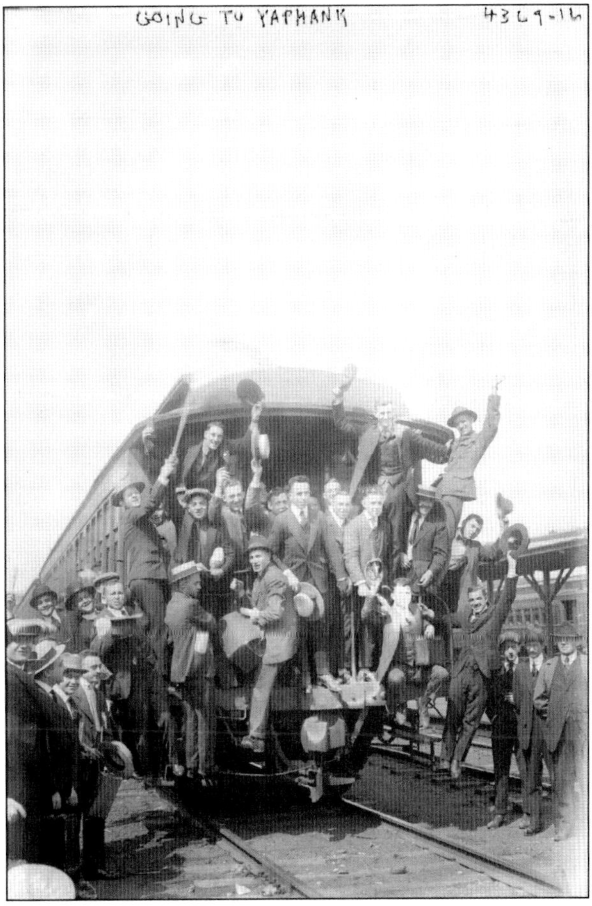

This enthusiastic group of drafted men from the New York metropolitan area is on its way to Camp Upton aboard a Long Island Rail Road train bound for Yaphank. According to the history of the 305th Infantry, they are carrying "a bundle of things mothers thought necessary for military life." The new training camp was carved out of the pine-barren wilderness in Suffolk County, New York. Once there, these recruits would become enlisted men.

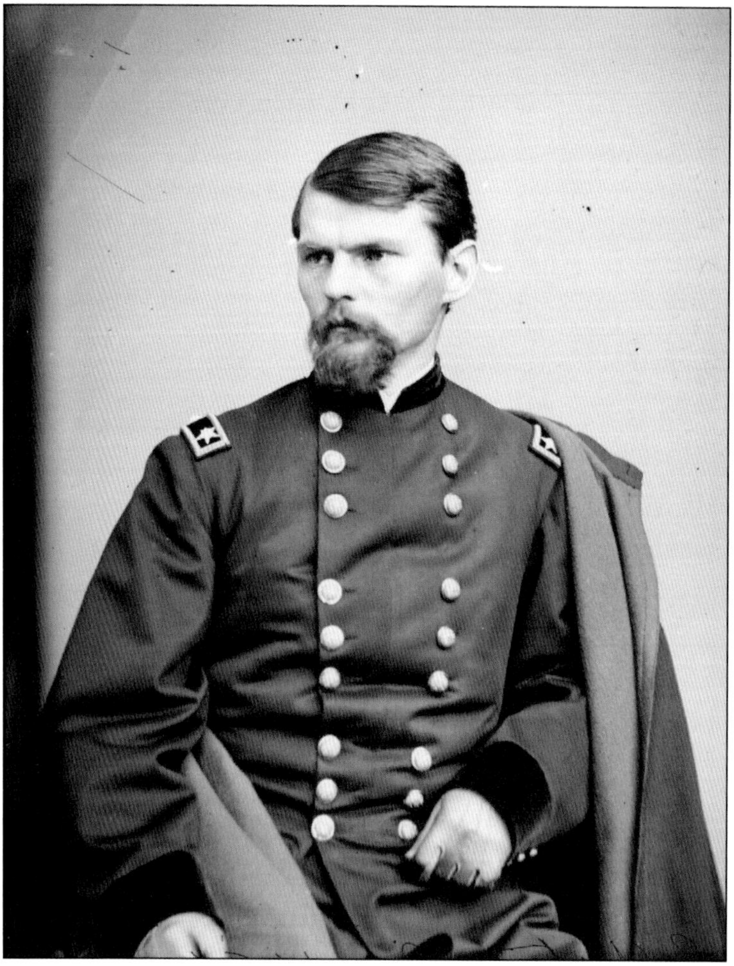

The cantonments were named for Army officers who were connected in some way to the geographic area they covered. Camp Upton was named for Maj. Gen. Emory Upton (1839–1881), who was born in Batavia, New York, and schooled at West Point (class of 1861). He served in the Army of the Potomac, was wounded at the First Battle of Bull Run, and commanded troops at Fredericksburg and Gettysburg. (Courtesy of the Library of Congress.)

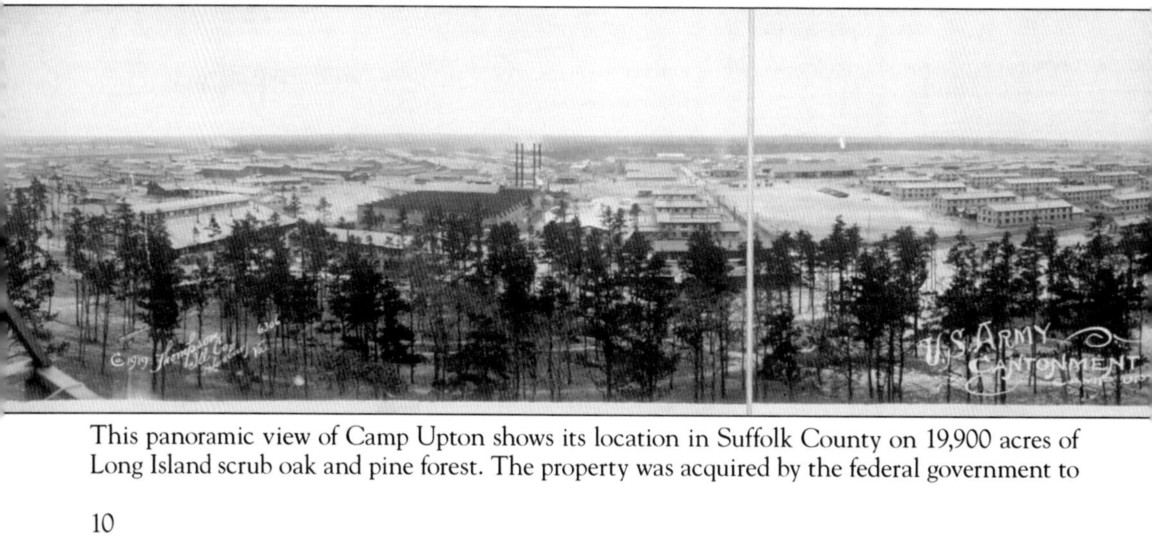

This panoramic view of Camp Upton shows its location in Suffolk County on 19,900 acres of Long Island scrub oak and pine forest. The property was acquired by the federal government to

Upton developed a tactic to successfully assault entrenched Confederate troops late in the war. After the war, he recommended changes in military training and organization. He served as commandant of the US Military Academy at West Point from 1870 to 1874. In 1868, he wrote *Infantry Tactics: Double and Single Rank, Adapted to American Topography and Improved Fire-arms*, followed by a revised edition in 1874 as well as *The Military Policy of the United States*, which was published posthumously in 1904. He is considered one of the most influential young reformers of the Army in the 19th century.

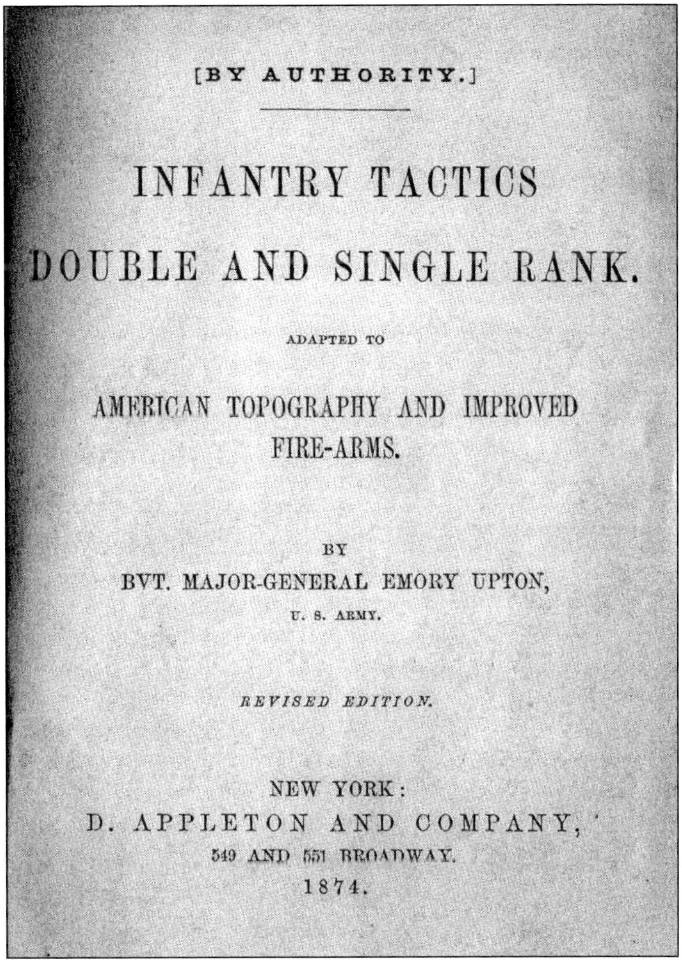

[BY AUTHORITY.]

INFANTRY TACTICS

DOUBLE AND SINGLE RANK.

ADAPTED TO

AMERICAN TOPOGRAPHY AND IMPROVED FIRE-ARMS.

BY

BVT. MAJOR-GENERAL EMORY UPTON,

U. S. ARMY.

REVISED EDITION.

NEW YORK:

D. APPLETON AND COMPANY,

549 AND 551 BROADWAY.

1874.

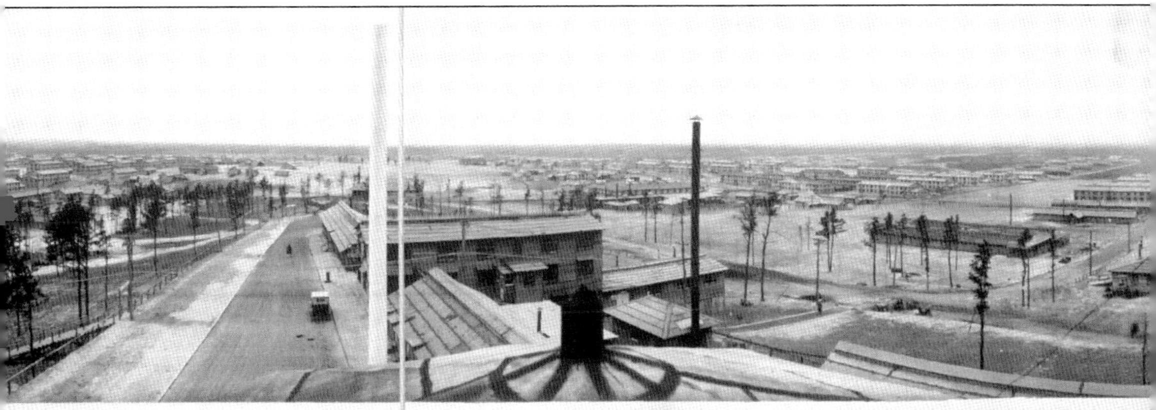

train 40,000 drafted men for the National Army at Camp Upton. They would form the 77th "Liberty" Division. (Courtesy of the Library of Congress.)

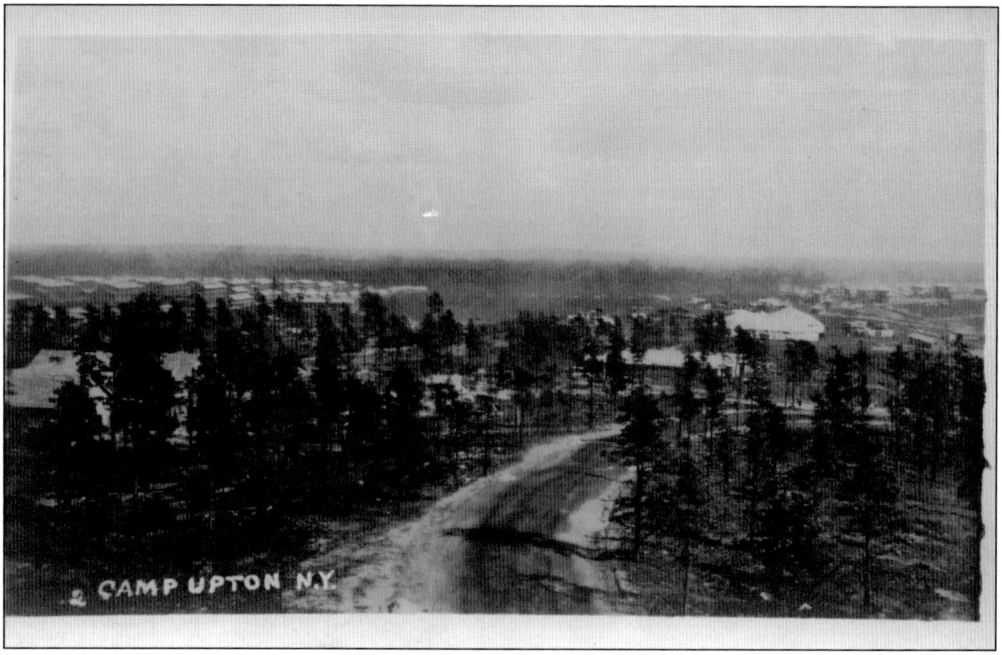

The Long Island Rail Road had extended east to the nearby village of Yaphank in 1844 as part of its route to Greenport. In June 1917, Col. Frank Lawton and Ralph Peters, president of the Long Island Rail Road, inspected the site in the wilderness at the suggestion of special agent Hal B. Fullerton. Easy access to the New York City metropolitan area by rail was one key element for locating the camp on eastern Long Island.

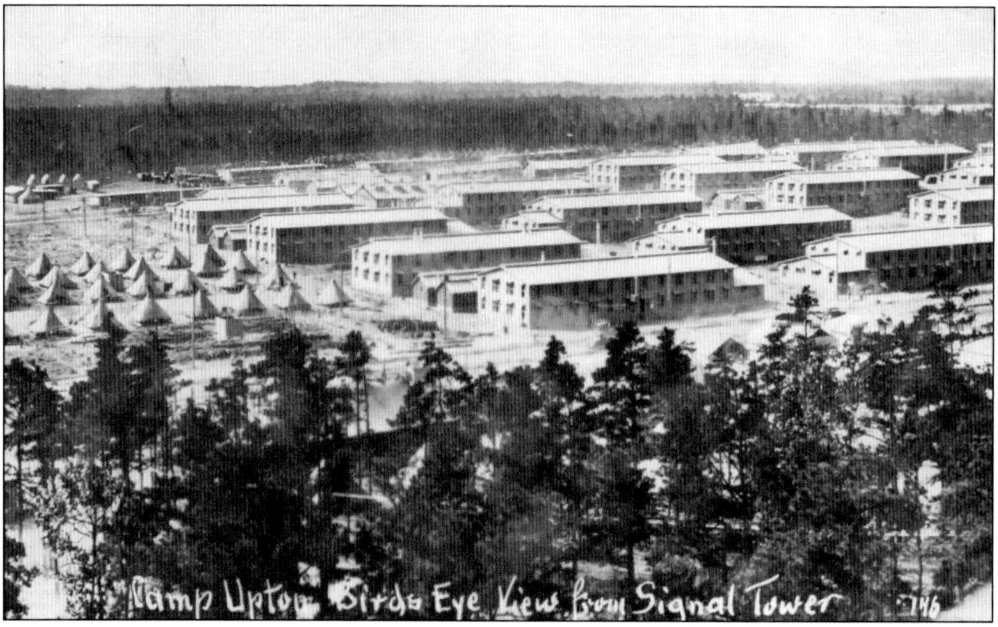

A bird's-eye view of the camp early in 1917 shows the tents that were in use as the barracks were being completed. The camp was the last of the 16 cantonments to be completed and was due to open in September, but its official opening was delayed to December 1917.

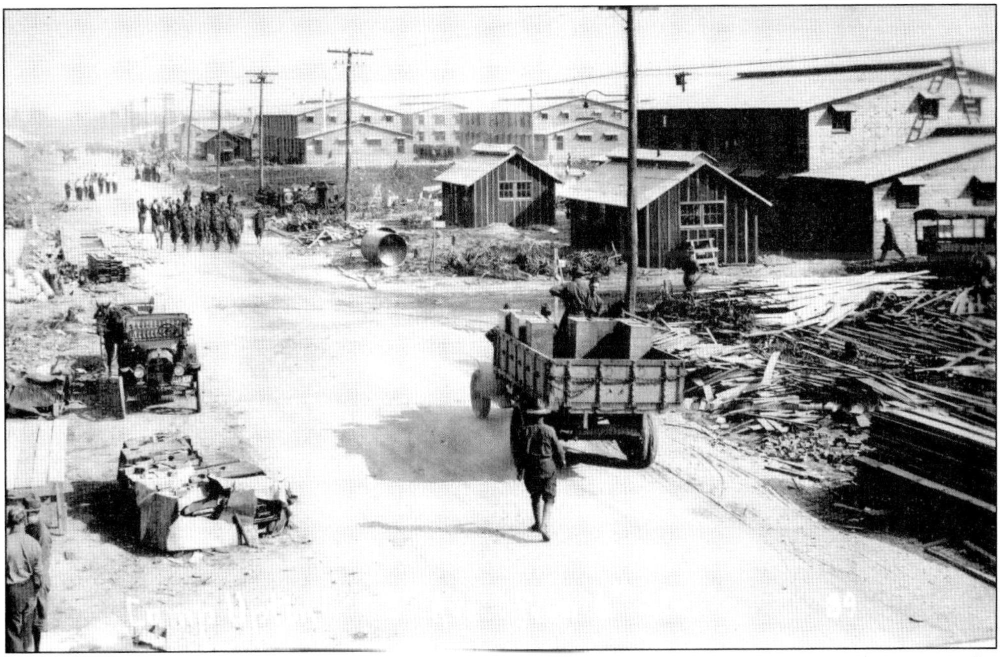

Camp Upton was built to quickly train 40,000 men from the New York metropolitan area for the new National Army. The contract to build the camp was granted to the Thompson Starrett Company of Wall Street. The first building was completed by August 1; by September, 60 barracks were ready.

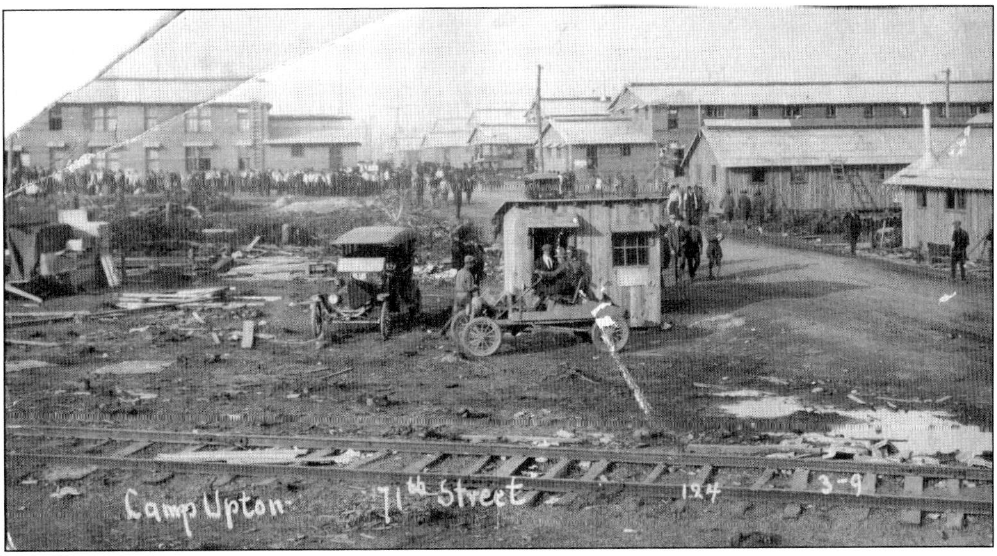

The contract called for the construction of more than 1,000 buildings, and work began in June 1917. The first soldiers were due to arrive that September. Early on, difficulties bringing supplies close to the construction sites contributed to the delay. Construction quartermaster J.O. Adler described it as looking "like a western mining town." The camp was not officially opened until December 1917.

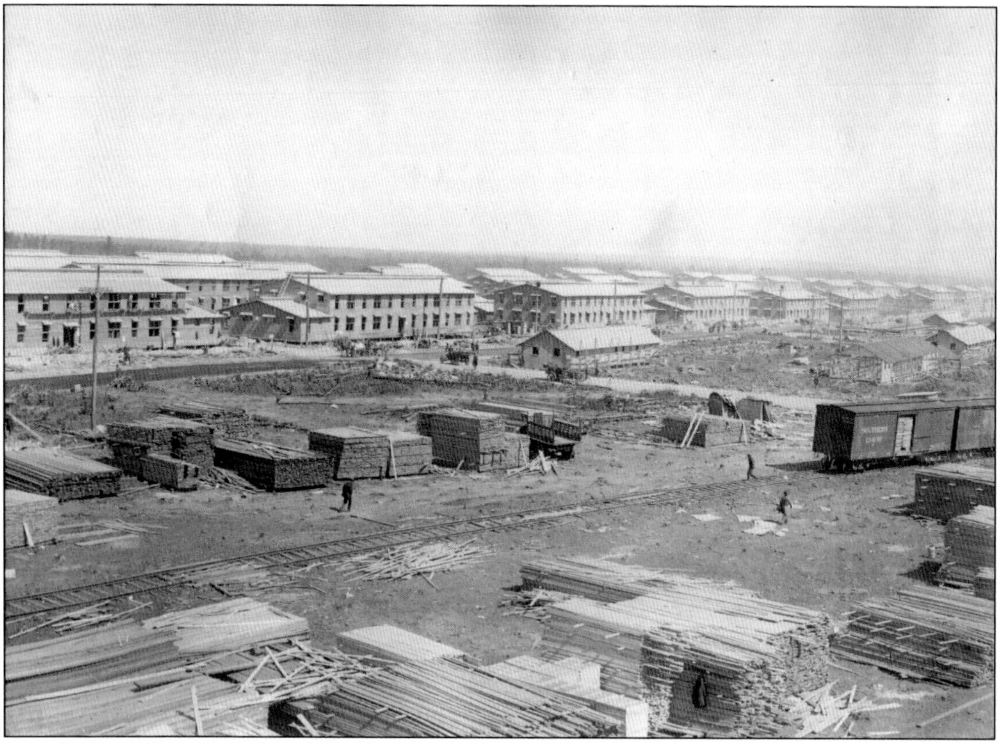

It took five million feet of lumber—low-grade unseasoned Southern pine—to build Camp Upton. Construction materials were received and stored adjacent to temporary railroad spurs, making the rapid distribution of supplies possible as well as the quick embarkation of troops later.

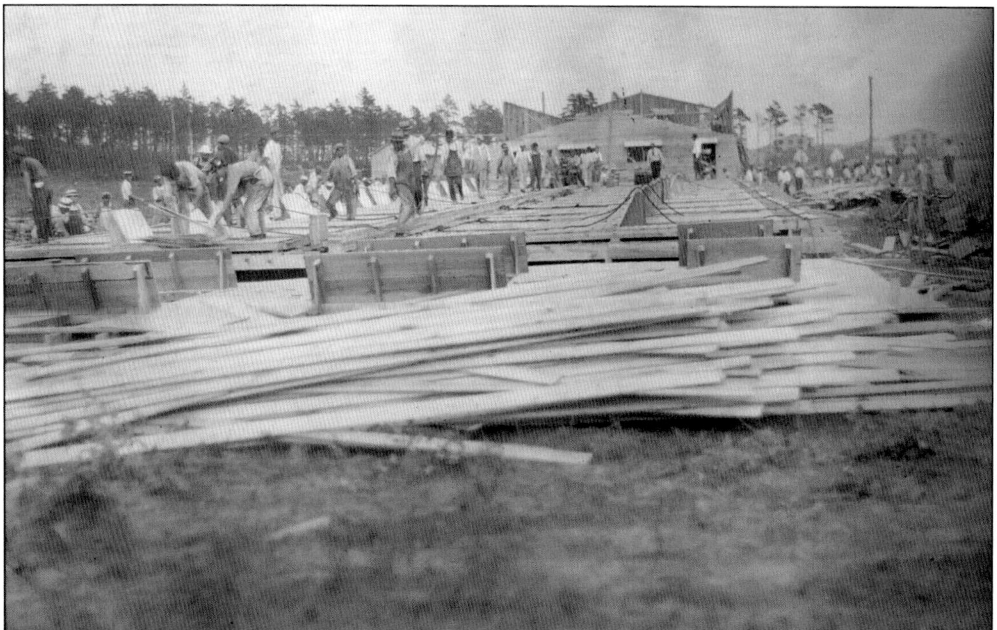

The walls of the barracks were 150 feet long and 2 stories high. They were constructed on the first floor and lifted into vertical position by several hundred workmen.

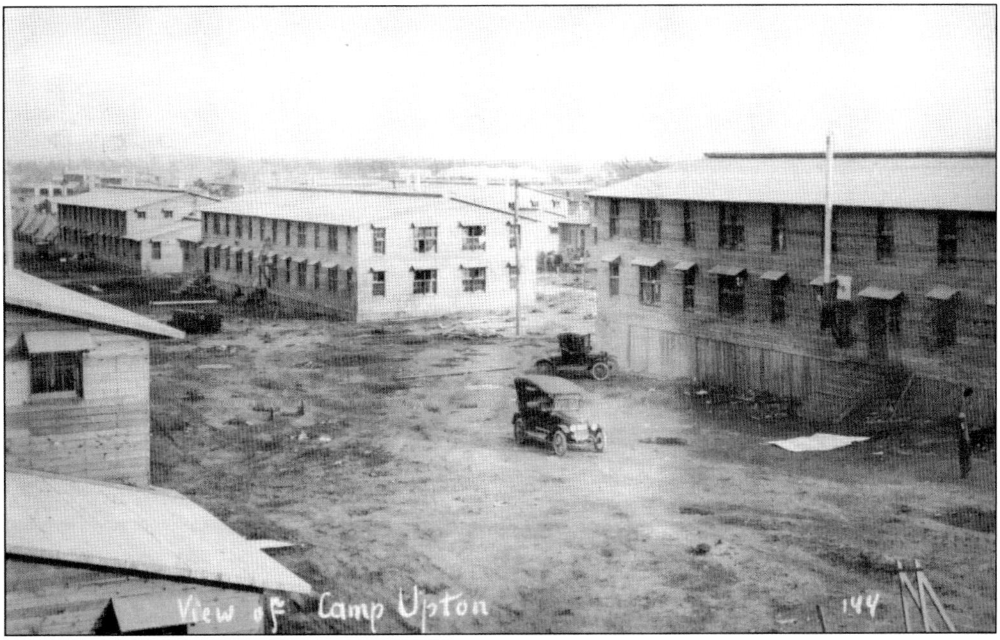

In a camp where all buildings were of pine, wide avenues separated the barracks of regiments in order to create fire breaks.

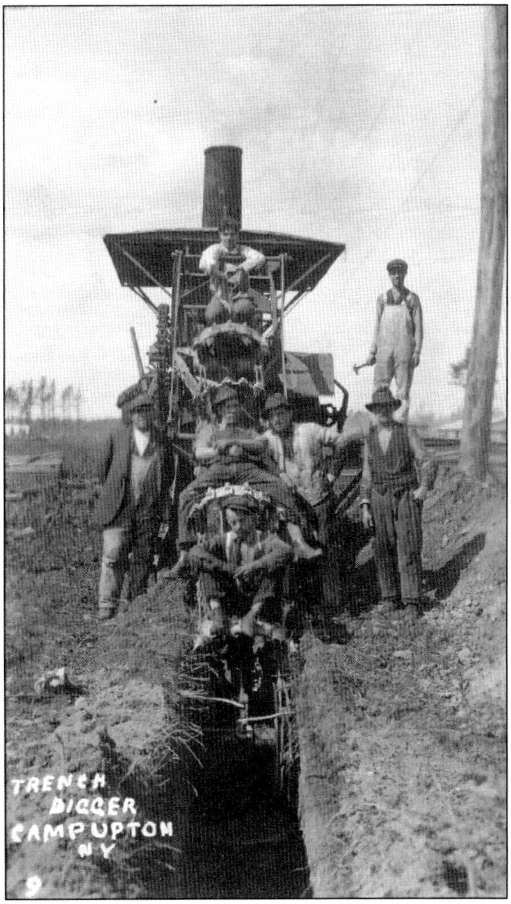

At first, 5,000 local men were hired at the camp for higher wages than farm work. But when local farmers complained that crops were not being harvested, laborers were brought in from New York City, and ultimately it took 15,000 men to build the camp. This crew is using a steam-powered trench digger to install water and sewer lines.

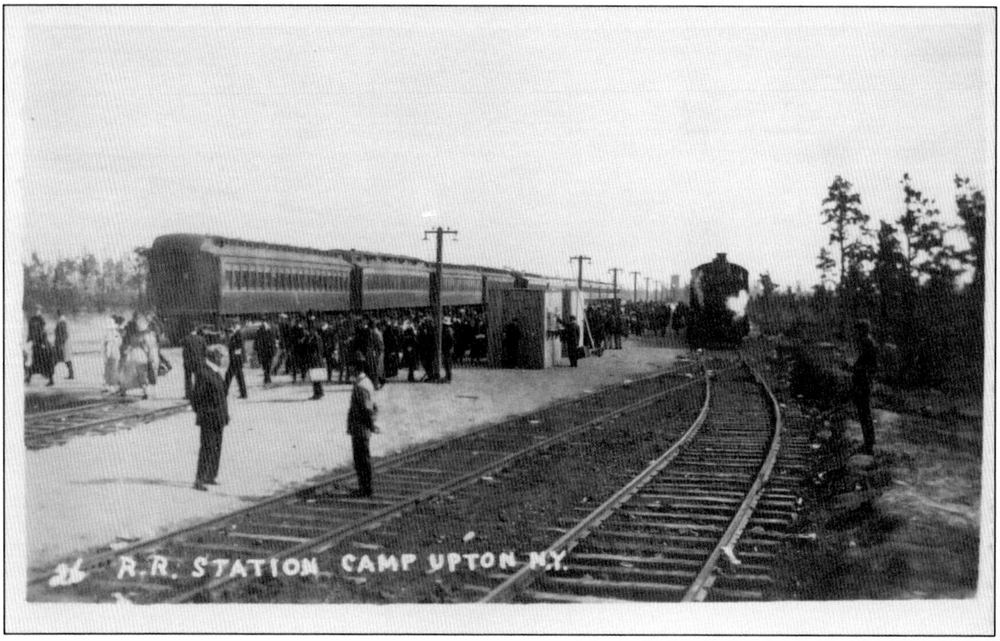

The Long Island Rail Road built a spur two miles out from the Yaphank station to the camp in 1917. Ultimately, there were 70 buildings at the terminal. Three trains traveled daily between Penn Station in New York City and Camp Upton and four on Sundays.

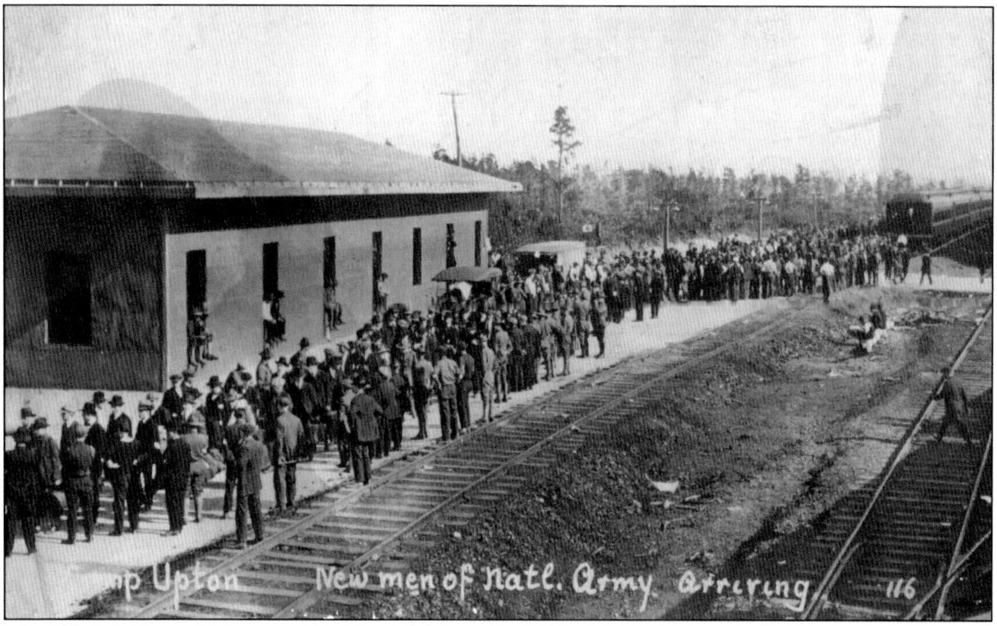

Not all arrivals at the camp went smoothly. An October 1917 postcard describes a train accident that occurred during the mad rush of visitors coming to the camp. Two soldiers were killed in the accident. (Courtesy of Paul Infranco.)

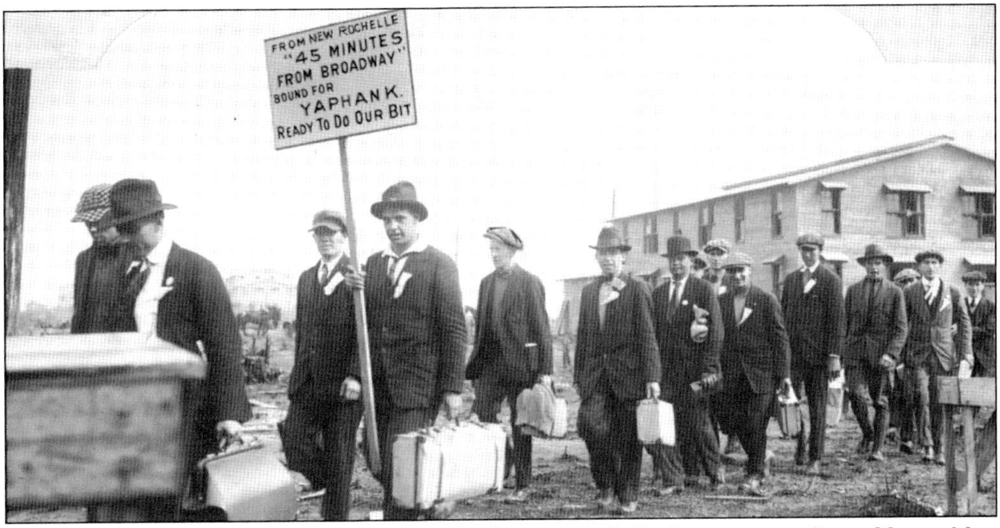

F.M. Stefano took this photograph, *Arrival of the New Rochelle Contingent at Camp Upton, New York*, in 1917. Draftees came primarily from New York City and the surrounding area, and the 77th was first known as the "Metropolitan Division." (Courtesy of the Library of Congress.)

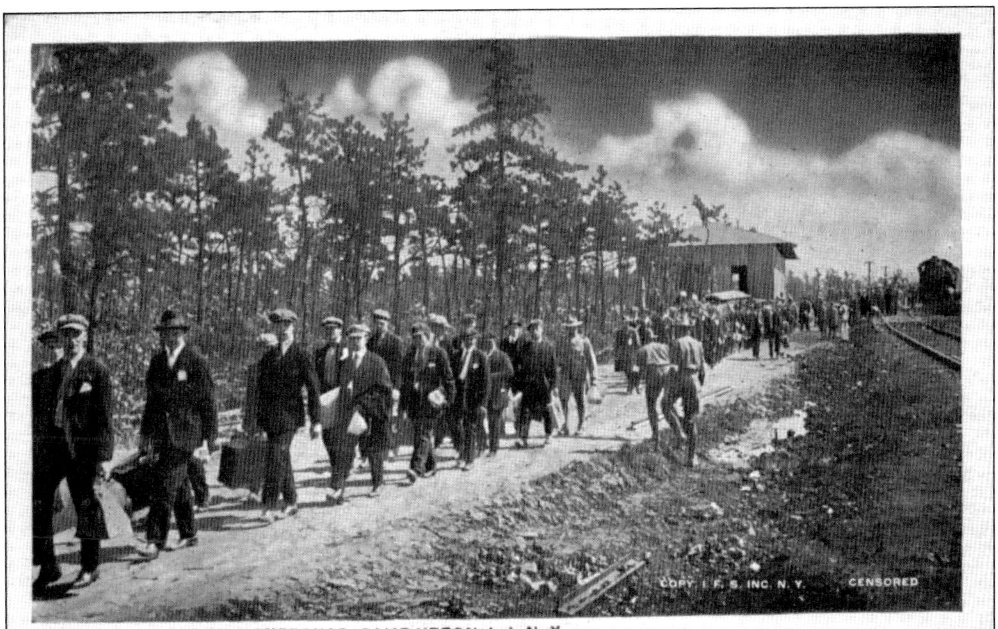

Entry into the war meant hundreds of thousands more soldiers were needed for the new National Army. On May 18, 1917, the Selective Service Act was passed, requiring all men 21 to 30 years old to register. Later, the range was expanded to 18 to 45 years of age.

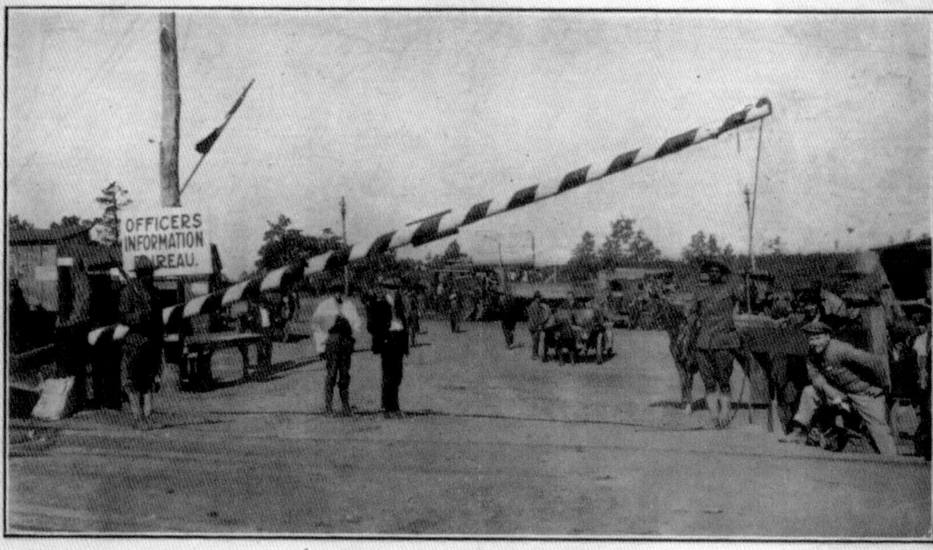

ENTRANCE TO CAMP UPTON

According to Thomas R. Bayles, the northern entrance to Camp Upton from Middle Country Road was the old Hay Road, which ran across the island to Mastic and south to the salt meadows of the Great South Bay. Salt hay was cut and used on farms for fertilizer. The Hay Road is now William Floyd Parkway.

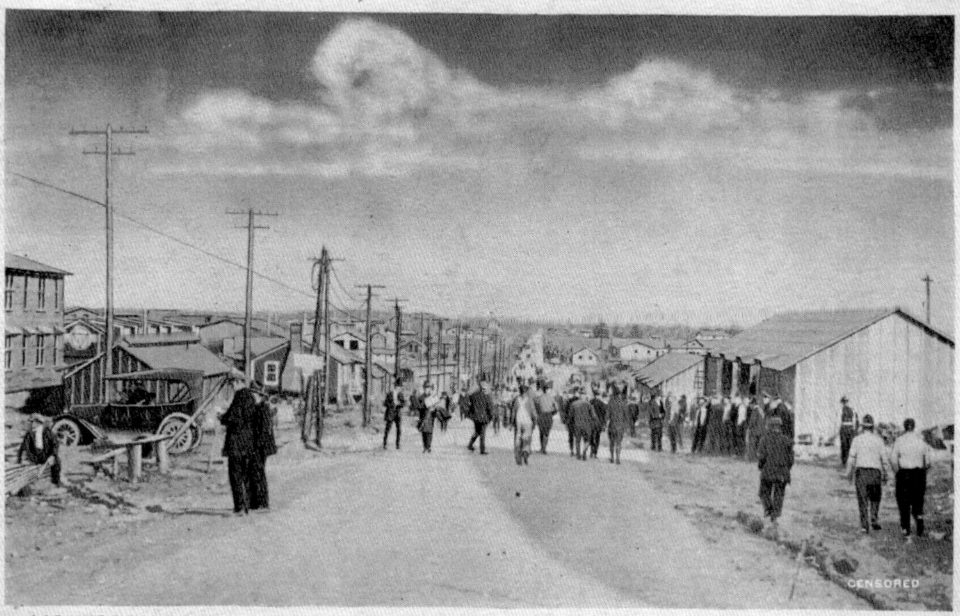

LOOKING DOWN SECOND AVENUE. CAMP UPTON, L. I., N. Y.

The streets of the camp were named for New York City streets, such as Fifth Avenue and Broadway, to give the men a sense of familiarity, as most of the draftees were from the metropolitan New York area.

Maj. Gen. J. Franklin Bell was commander of the US Army's Eastern Department with headquarters at Fort Jay on Governors Island in New York City in 1917. A West Point graduate, he was an officer in the 7th Cavalry during the Indian Wars, earned the Medal of Honor during the Philippine-American War, and served as chief of the Army general staff under Presidents Roosevelt and Taft. His responsibilities included supervision of the Plattsburgh Officers Training Camp.

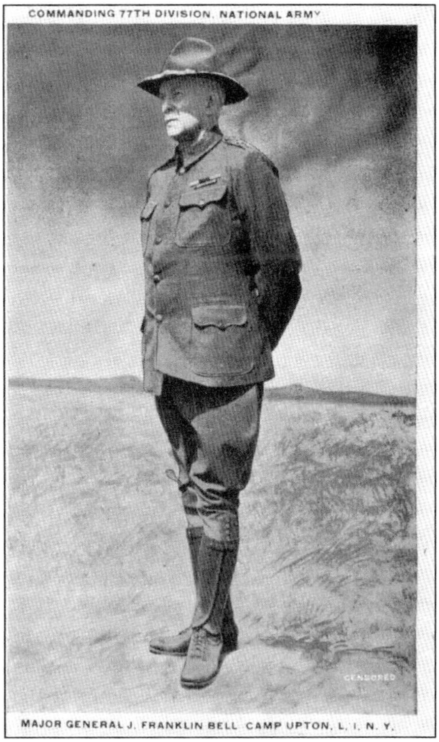

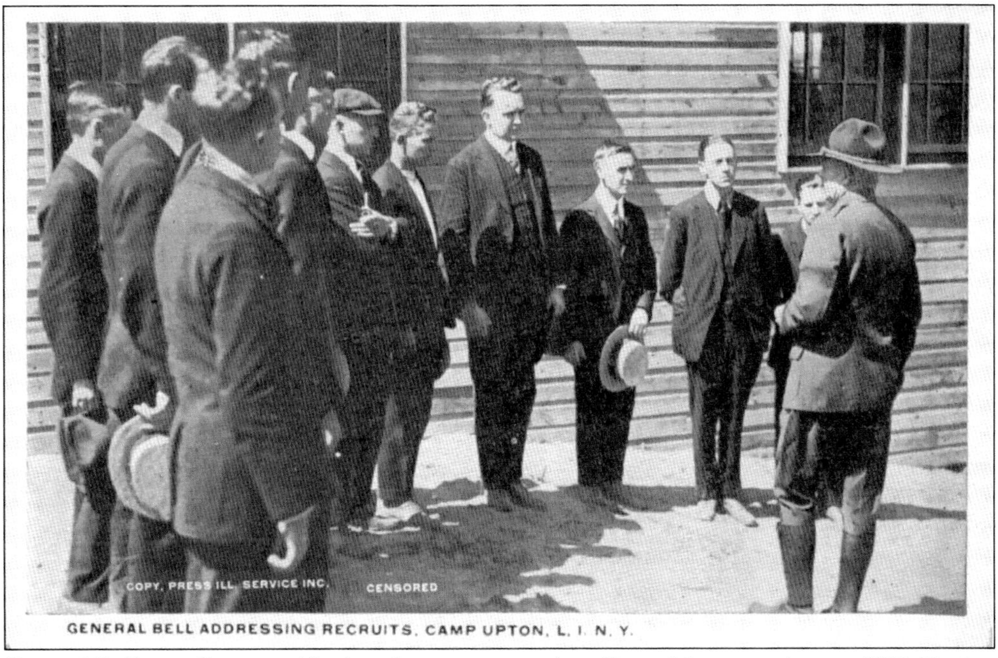

General Bell was 61 years old when he was assigned to Camp Upton in August 1917. During training, General Bell was known for walking around in the camp and speaking to individual troops and officers, often complimenting them. His health prevented him from going to France with his division, and he was replaced by Gen. Evan Johnson in April 1918. He died at Governors Island 18 months later.

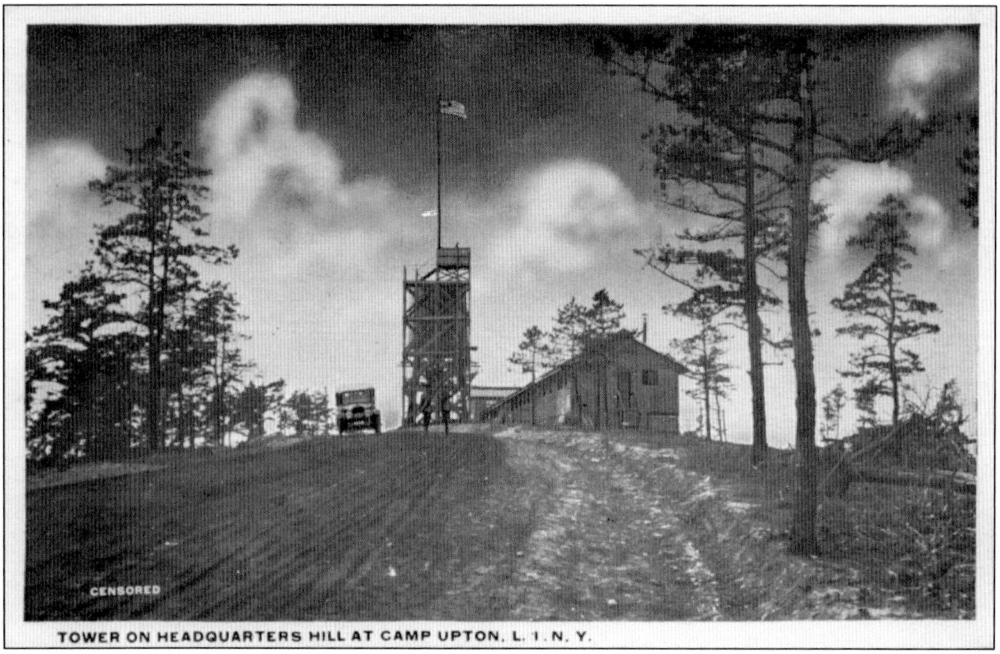

Tower Hill was on the ridge running east to west in the center of the camp. Maj. Gen. J. Franklin Bell selected the site for his headquarters. From the top of the hill, General Bell had a full view of Camp Upton and its surroundings.

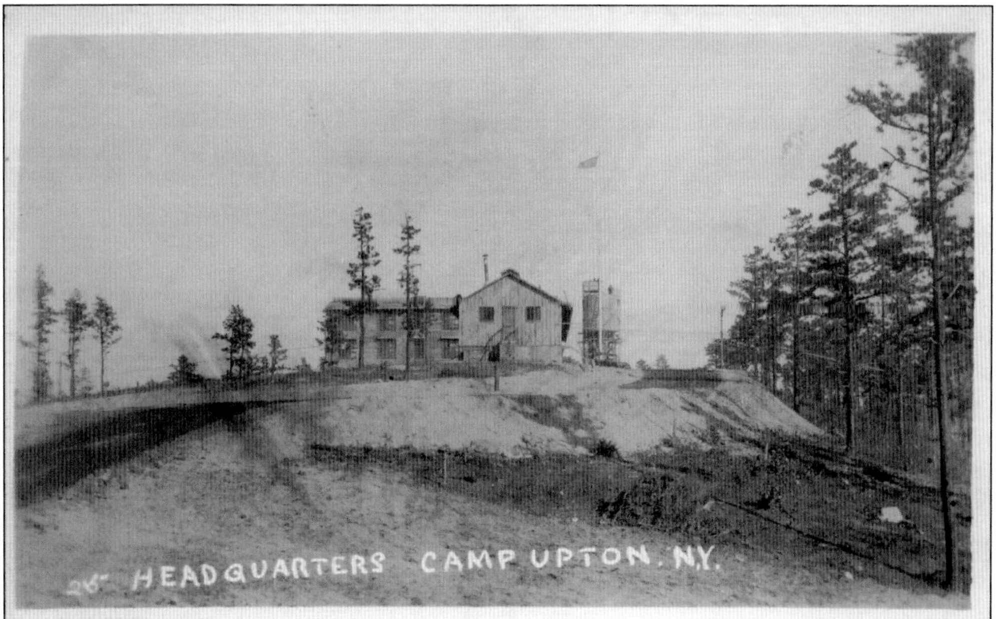

Headquarters Hill was a popular spot for visitors. Soldiers could bring guests on tours that included climbing the water tower.

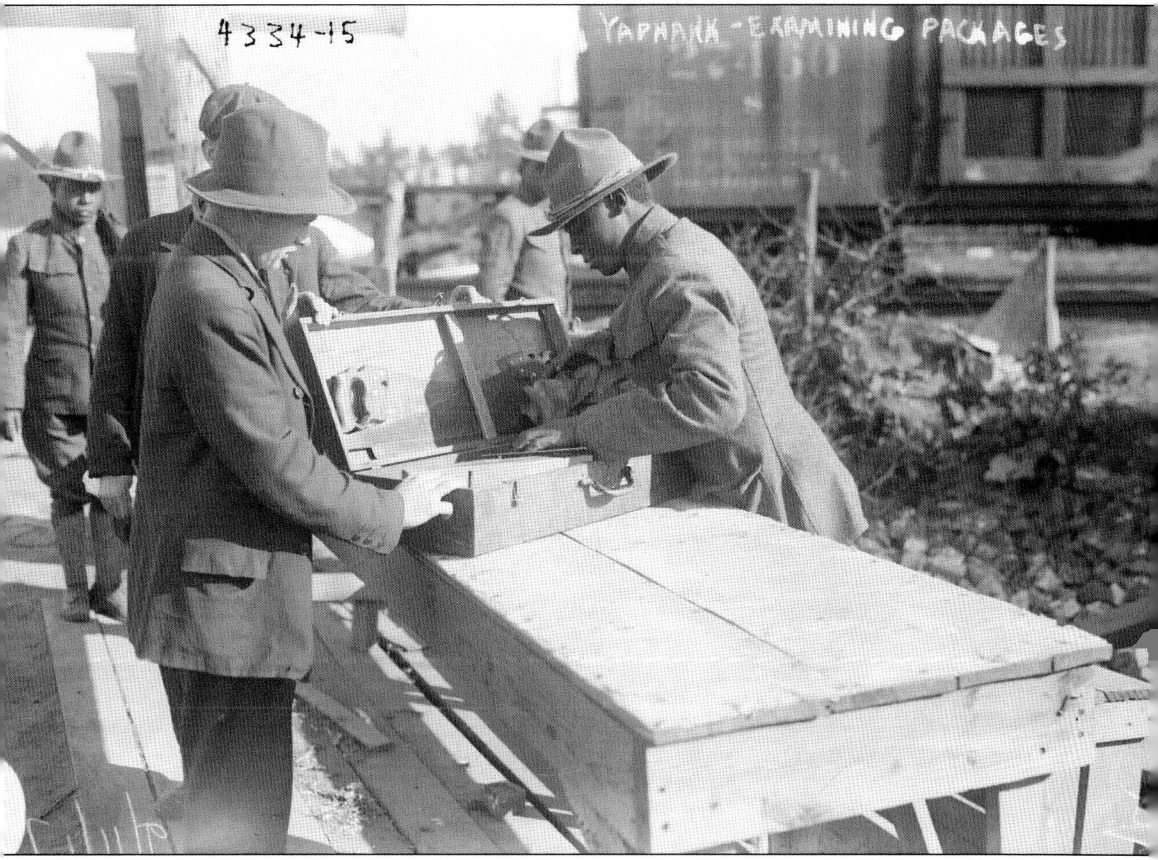

The 15th New York National Guard, a regiment of black soldiers from New York City, was called into federal service as the 369th Infantry on July 25, 1917. It completed basic training at Camp Whitman, and its three battalions were assigned to guard rail lines, construction sites, and other camps in New York. One of its battalions was ordered to Camp Upton. There, the black soldiers lived in separate barracks but shared the YMCA and other recreational facilities with the recruits. Their duties included guarding the gates and perimeter of the camp. They were also responsible for performing the unpleasant task of inspecting packages coming into the camp and removing alcohol and weapons. On October 8, 1917, the entire regiment traveled to Spartanburg, South Carolina, for combat training. This regiment was later sent to France and assigned to the French army, where its outstanding performance in combat earned the soldiers the nickname "Harlem Hell-Fighters." (Courtesy of the Library of Congress.)

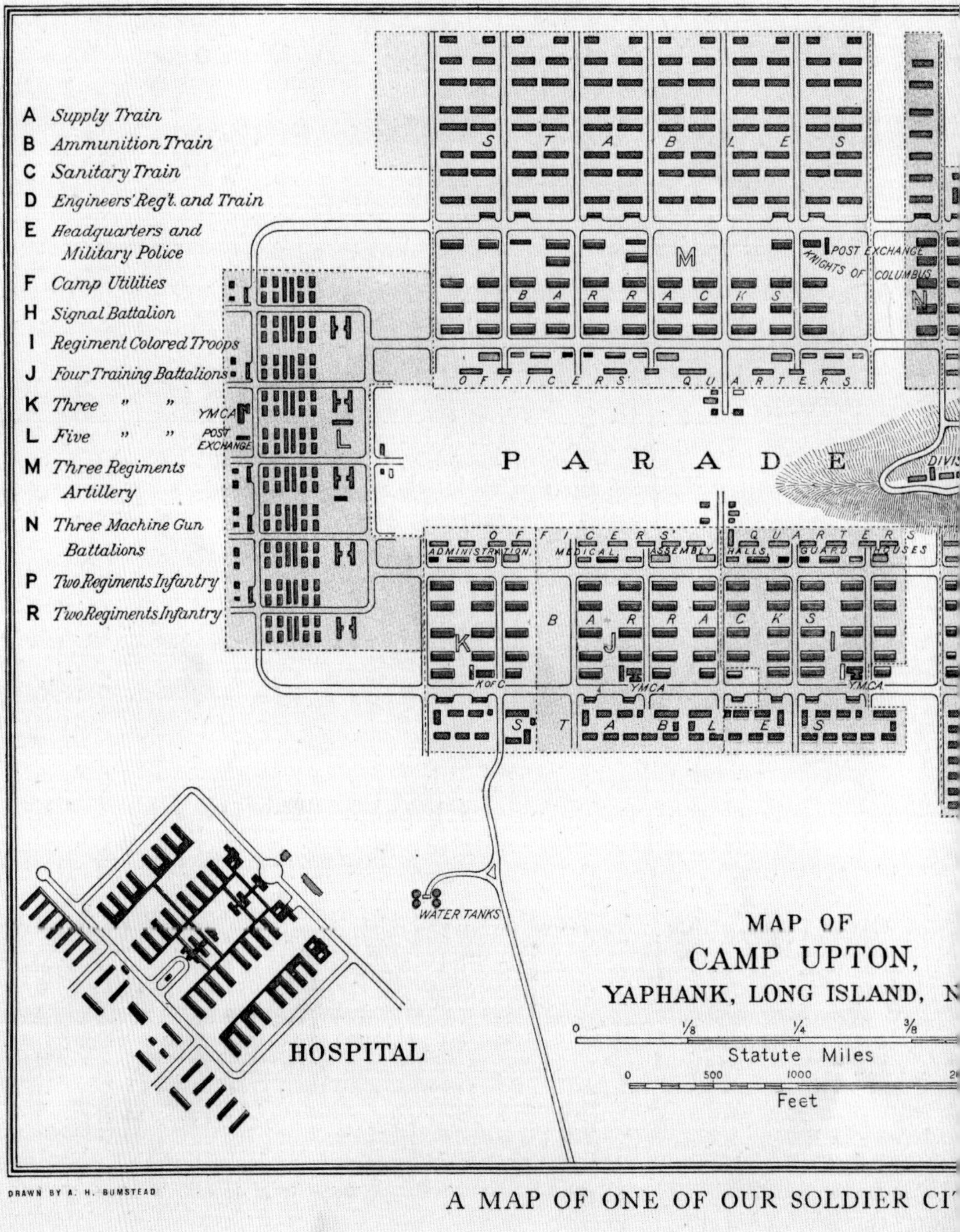

A MAP OF ONE OF OUR SOLDIER CI

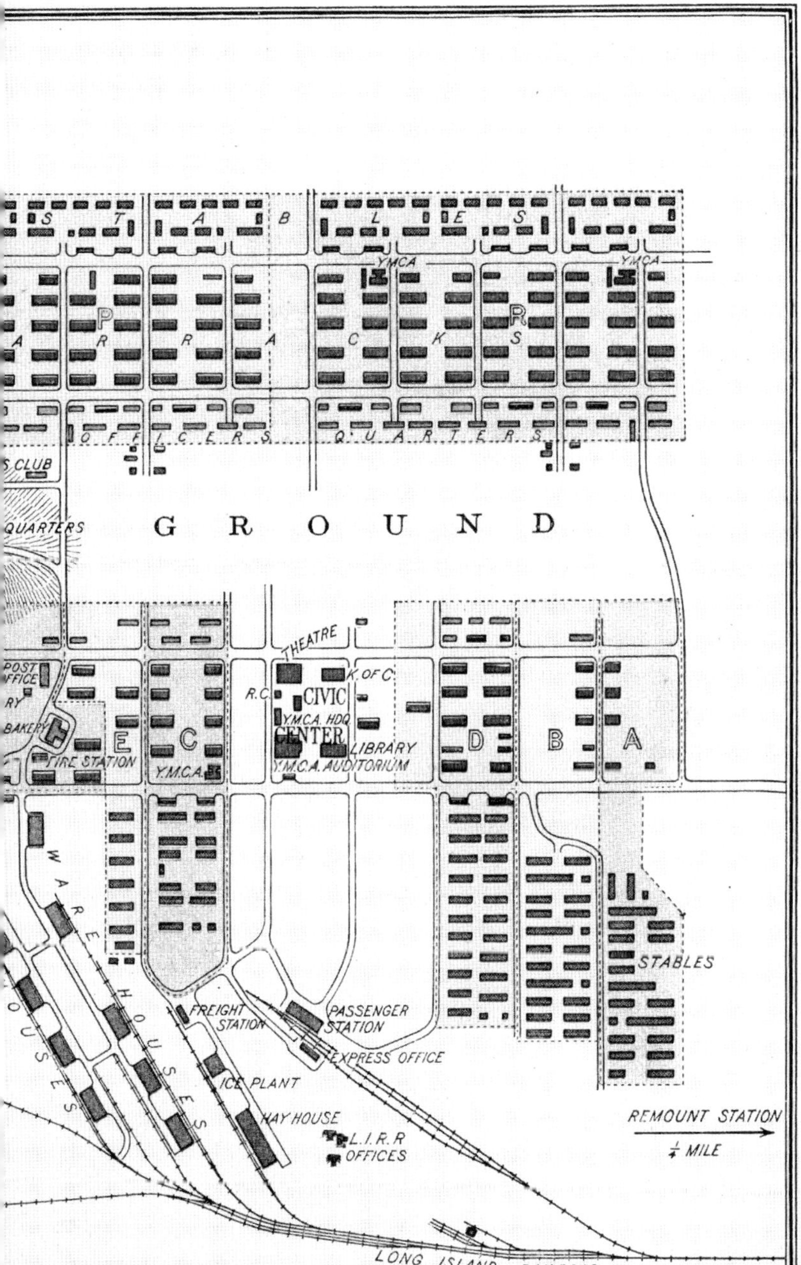

—See Pages 425 and 426

The map of Camp Upton seen here, drawn by A.H. Bumstead, appeared in a special double issue of *National Geographic* magazine (November–December 1917) dedicated to the war effort. "The standardized shape was to be similar to the letter 'U', with division headquarters to be situated at the bend of the letter and the soldiers' houses to spread regularly through the branches," wrote W.J. Showalter in the magazine. Each branch equaled about two and a half miles, so that all units were close together and near headquarters. The design of the National Army camps, or "soldier cities" as they were called, was meant to follow the same plan in each location. "The first view of one of these camps must impress the soldier with the fundamental element of military life—order through efficiency."

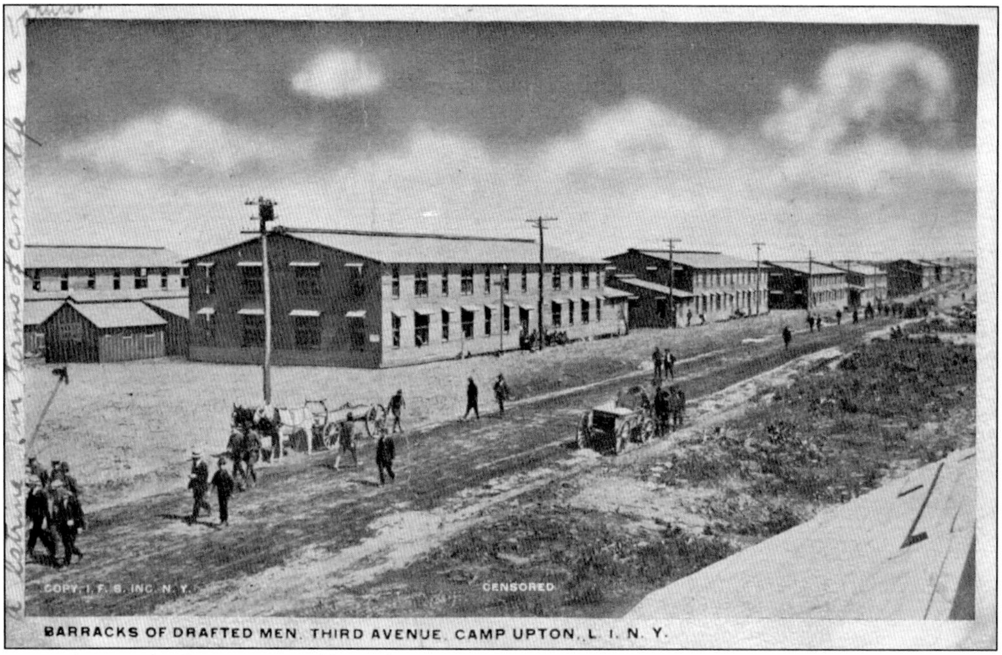

Barracks were built in groups of four, the number necessary to house a battalion of four companies, one third of an infantry regiment. Each barrack was designed to house a company of infantry for a battery of artillery, about 200 men. A latrine was located behind each barracks building and a company street in front. The recruit who sent this card is rapidly learning the military vocabulary, as he notes "a latrine—in terms of civil life a bathroom."

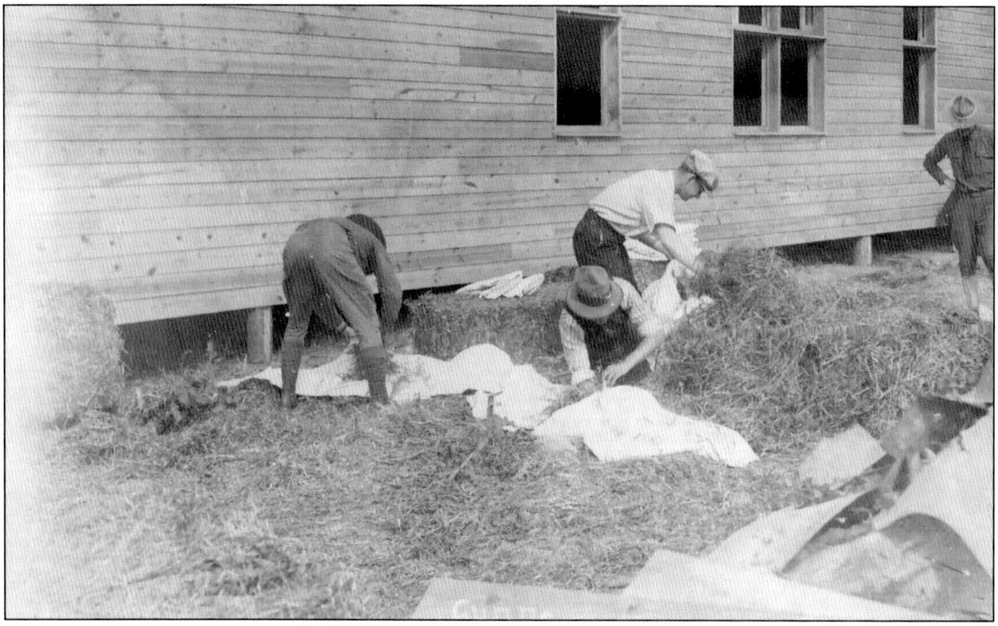

As the recruits arrived, each man was given a mess kit, two or three blankets, and a bed sack. The soldiers shown here are stuffing the sacks with straw to make mattresses for their cots. Soldiers commented to family members "there are plenty of blankets." All mattresses and bedding were to be aired thoroughly each morning by hanging them out the windows.

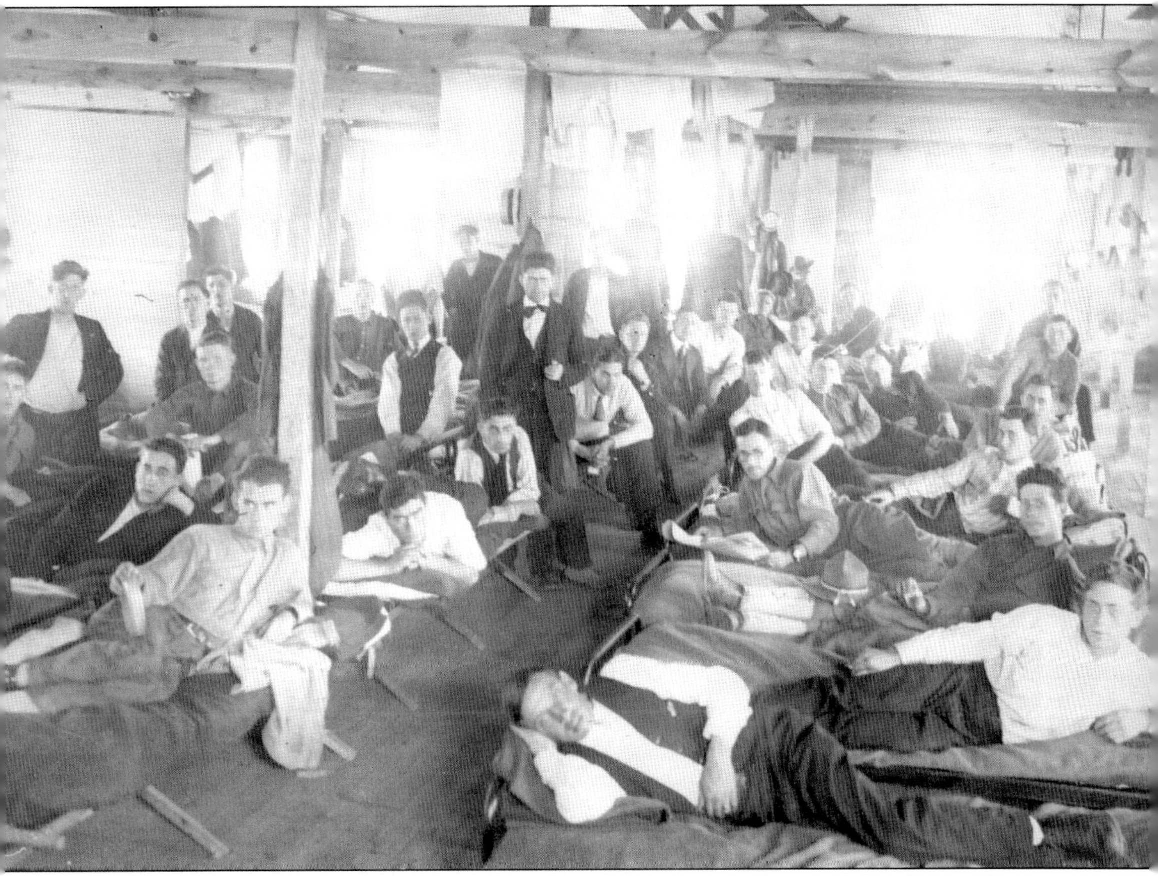

When the first recruits arrived, they spent their days getting to know each other in the new soldier city. For most draftees, this was their first experience living in close quarters with men from diverse backgrounds. At least 26 different languages were spoken by the recruits. Many could not speak English and were not yet full-fledged citizens.

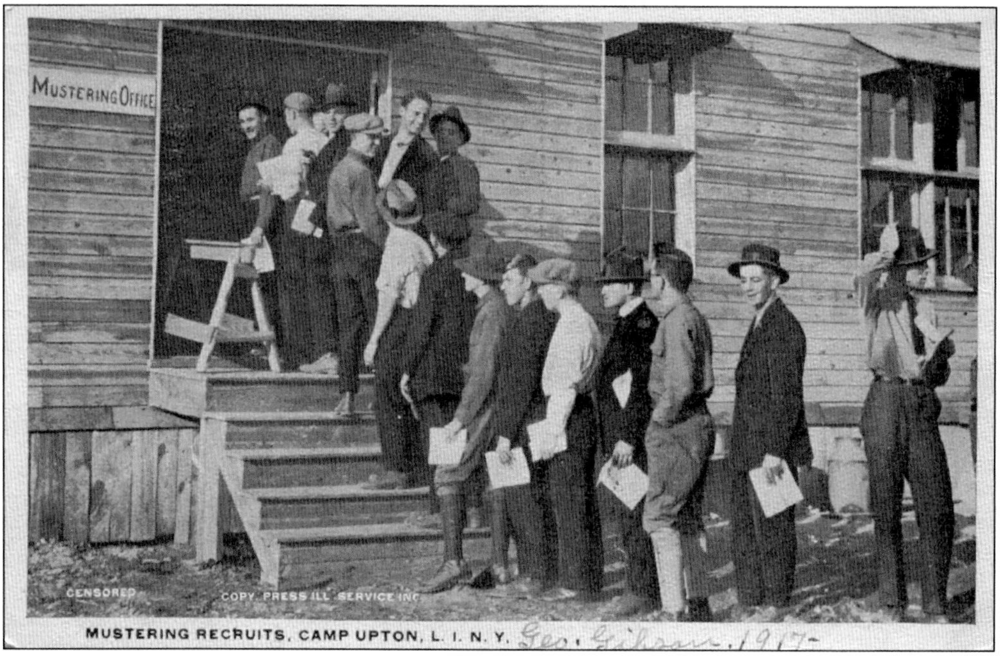

The first 2,200 recruits arrived on September 10, 1917, before the camp was completed. The mustering office was the first stop. It was open 24 hours a day, 7 days a week, admitting 2,000 soldiers a day. The basic military training was intended to last 16 weeks.

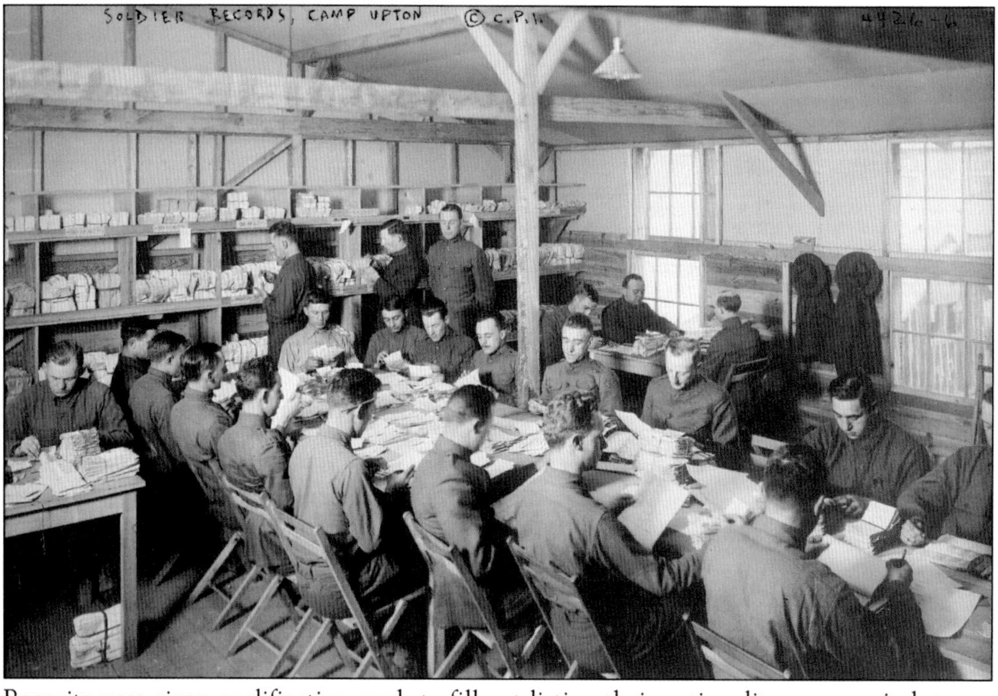

Recruits were given qualification cards to fill out listing their nationality, age, marital status, education, work experience, and what the soldiers called "their entire pedigree." Officers would then review the information and designate assignments. Those draftees who were discovered to have children were sent home. (Courtesy of the Library of Congress.)

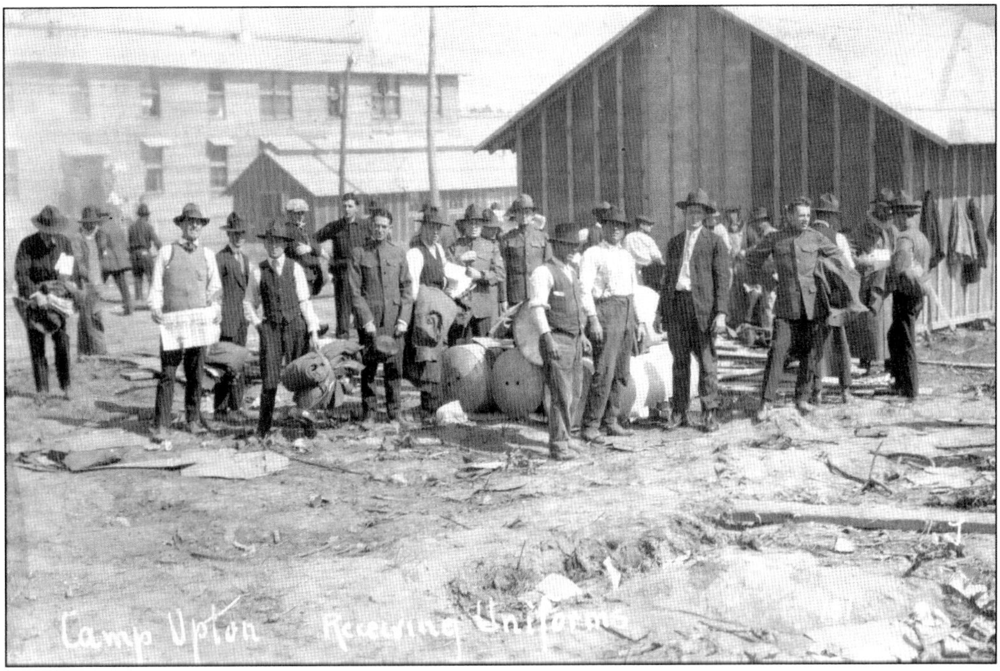

Uniforms were distributed in the first few days after arrival at camp. The uniform consisted of a jacket, pants, leggings, boots, several pairs of stockings and underwear, and a flat-brimmed campaign hat. Winter coats and overseas caps came later.

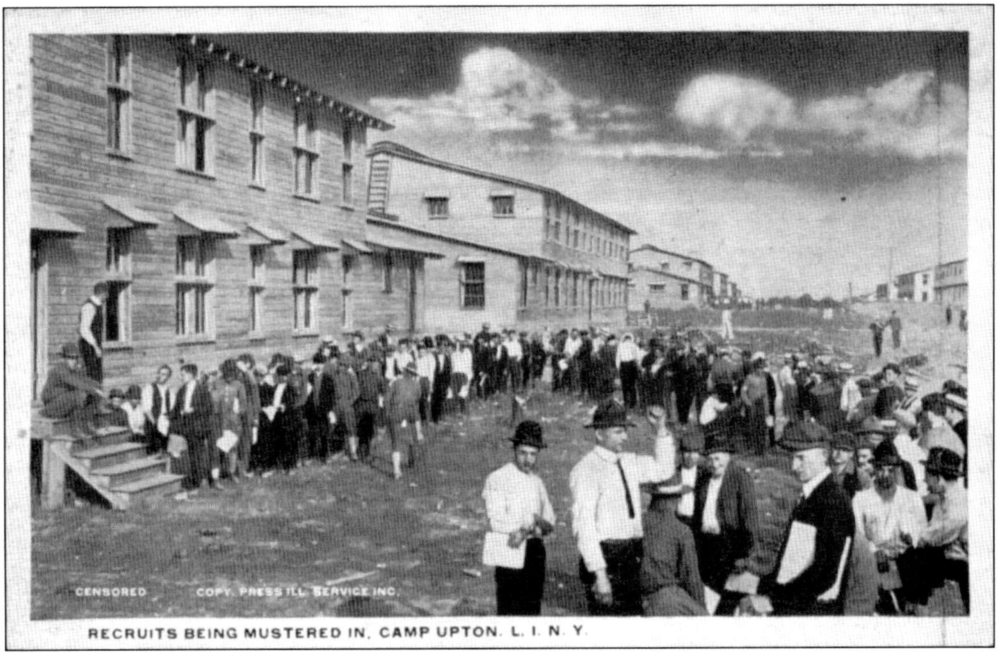

Recruits completely overwhelmed the unfinished camp in September 1917. By that date, only one third of the camp was ready. Barracks buildings were shared by several units until more were completed. These men look ragged because they were told to arrive in "throw away clothes," as they would not need them once they received uniforms.

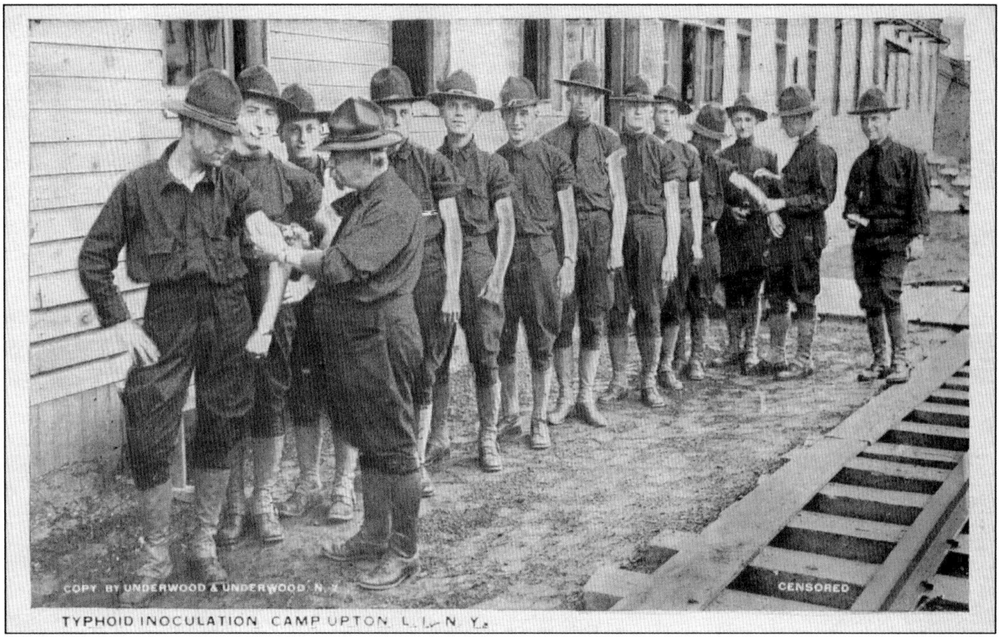

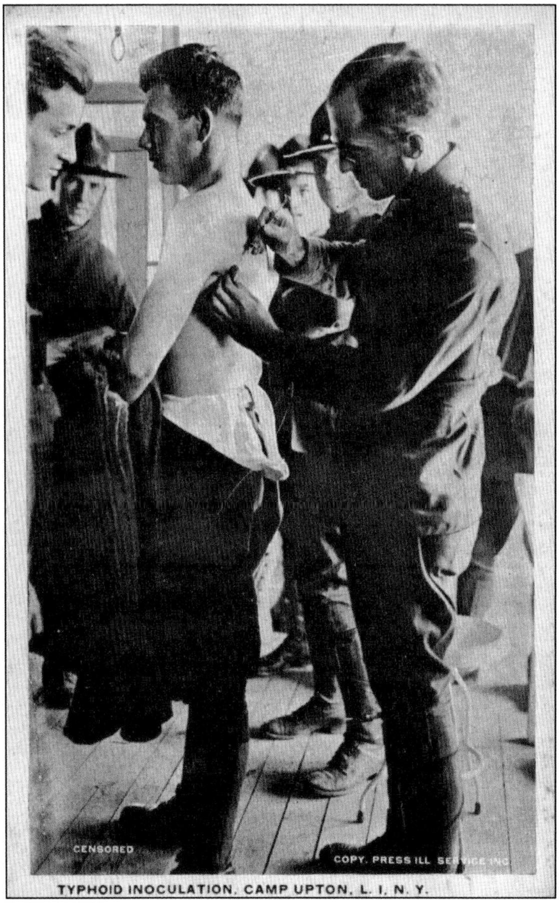

Soldiers received several inoculations for typhoid, tuberculosis, and diphtheria. However, there was no vaccine for the flu, and an epidemic in September 1918 severely impacted Camp Upton. Two thousand soldiers were treated for the disease at the base hospital, and nearly 500 died of pneumonia. "Soldiers from the 319th Engineers who came through from Camp Fremont in California on their way to France contracted the flu, and took it to Europe," wrote Barbara Wilcox, in her book *World War I Army Training by San Francisco Bay*.

A diphtheria outbreak in December 1917 resulted in a period of quarantine at the camp. Diphtheria is a highly contagious blood disease.

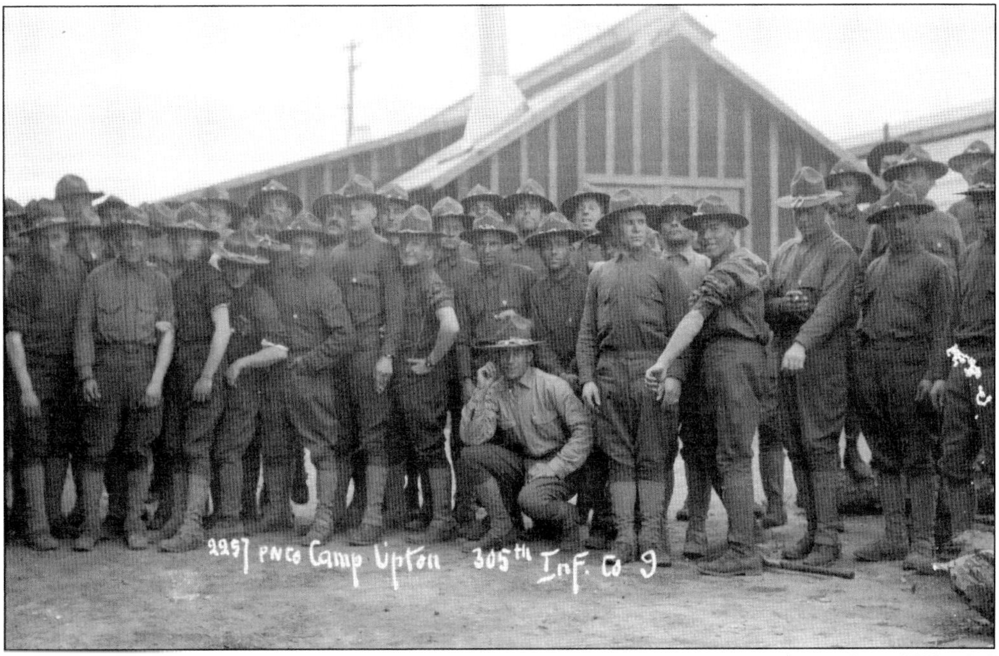

As new recruits arrived, those who had already been initiated would warn the rookies, "Wait 'til you see the needle." Soldiers received seven inoculations and two vaccinations.

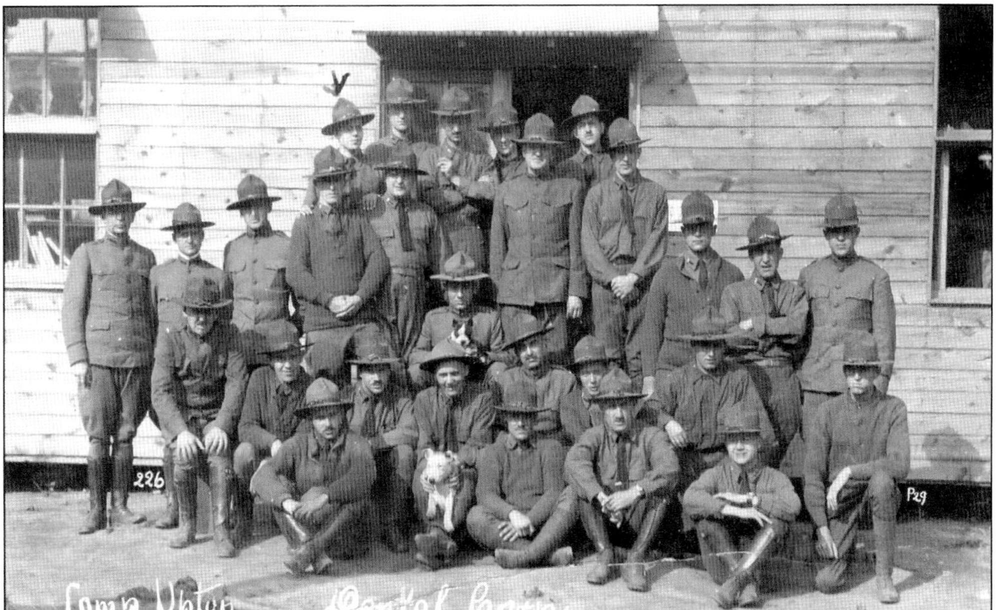

The US Army's first dental officer was commissioned in 1911. The importance of dental health was recognized in 1917, and every cantonment had a contingent of about 30 dental officers. Most of the men of the 77th Division had never seen a dentist before arriving at Camp Upton.

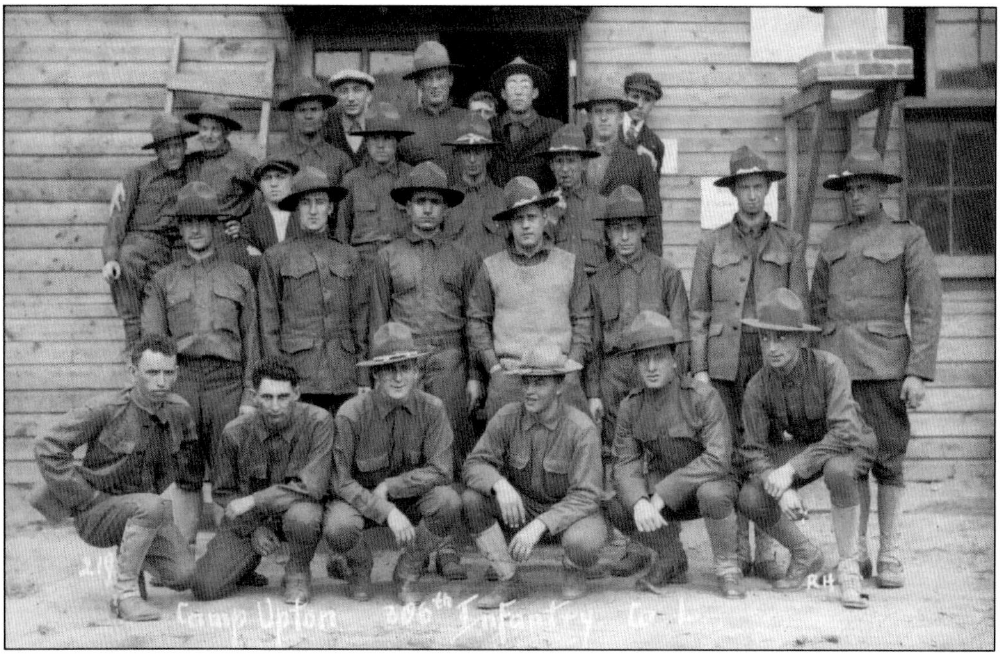

When war was declared, America had not been involved in a major war for 50 years and had not maintained an army of the size that would be required to fight the Germans. The US Army was ranked 17th in the world. Many young men who had only known peace signed up out of patriotism and love of their country. Several of the men pictured here are recent arrivals who have not yet received uniforms. They were called "rookies" by the men who had arrived earlier.

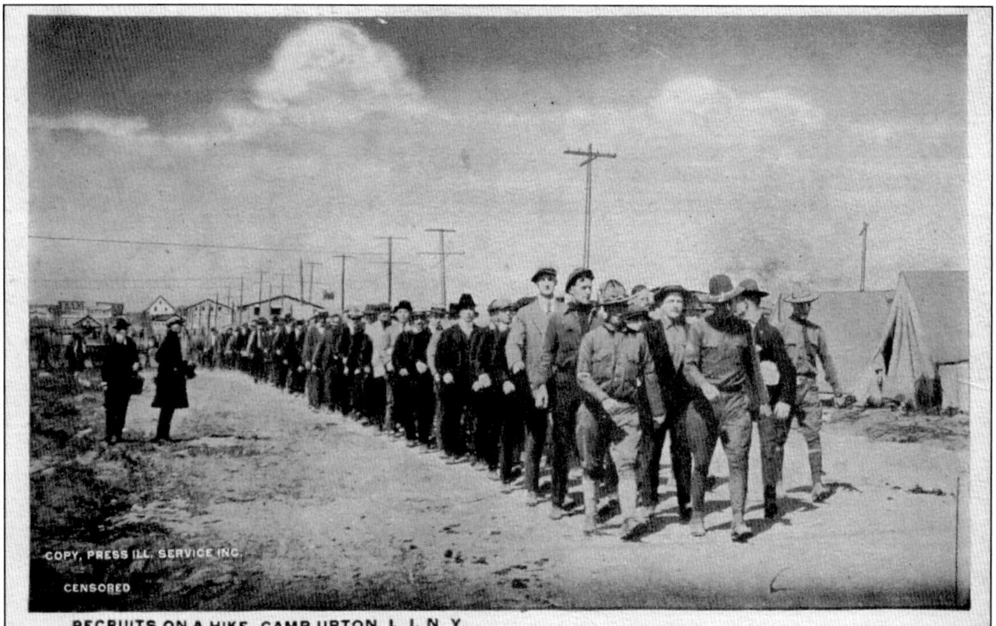

In the early days at the camp, officers who had completed the course at the officer training camp at Plattsburg, New York, began the process of turning draftees into soldiers. One officer complained "we are drilling eight hours a day."

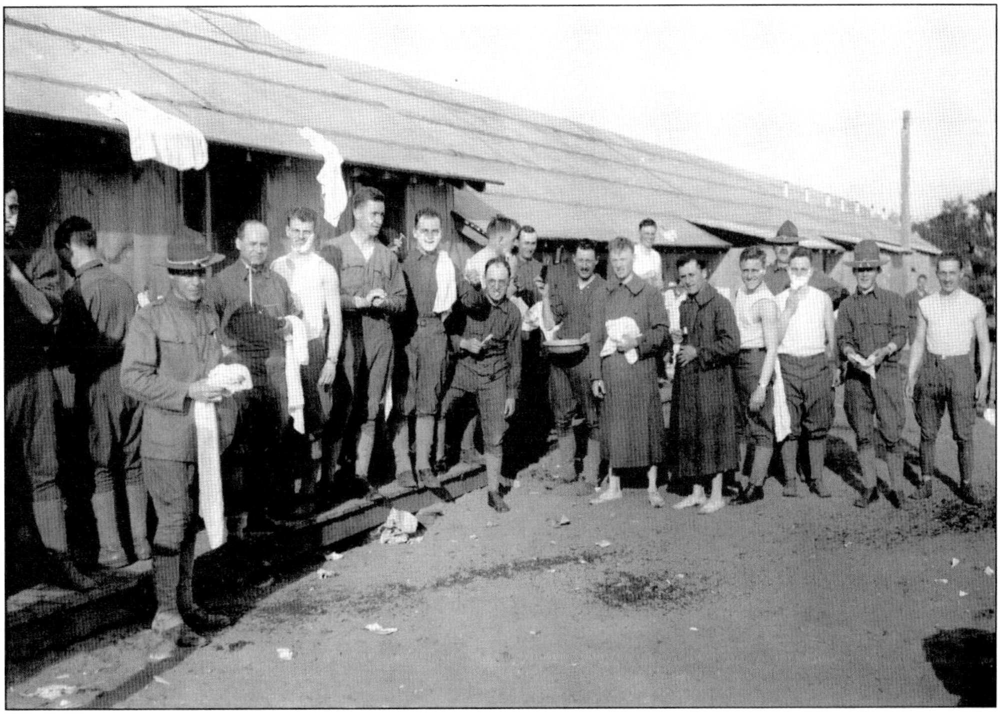

This unusual photograph shows recruits sprucing up at the latrine facility for daily formation.

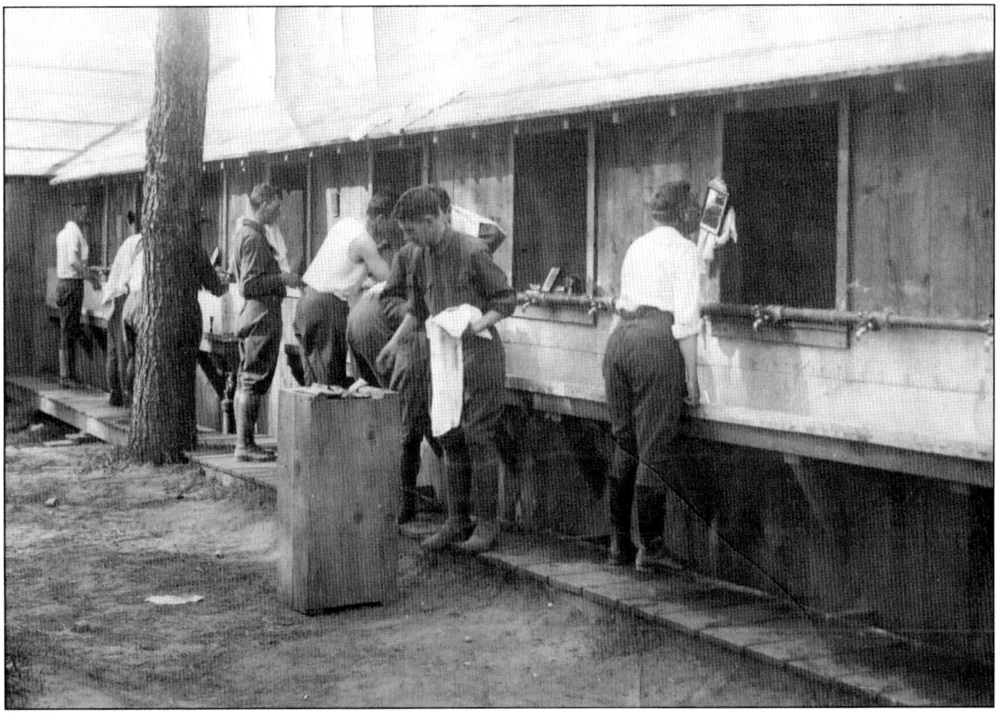

The camp buildings were outfitted with running water, and showers were expected to be taken in the shower house on a daily basis, which for many was a shock. The camp featured electricity from the Northport plant of the Long Island Lighting Company.

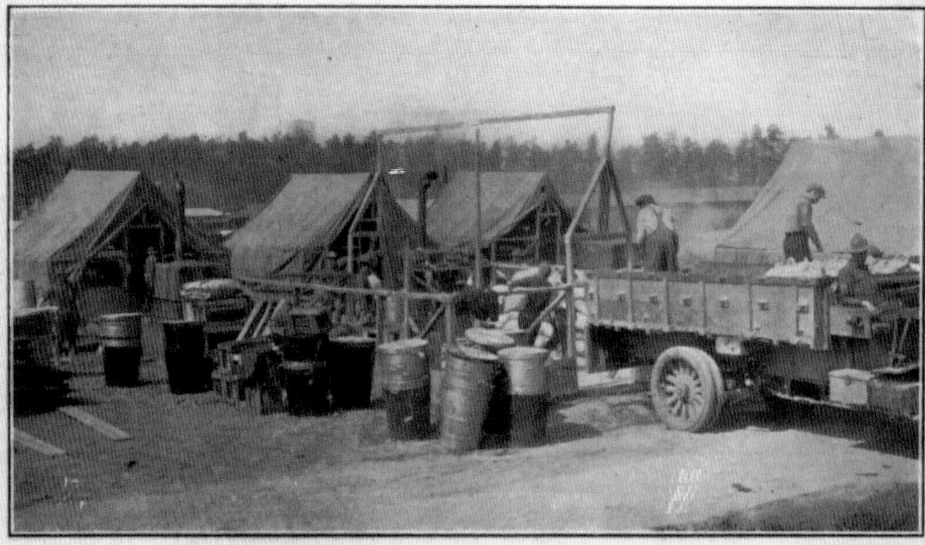

BRINGING IN SUPPLIES — CAMP UPTON

Providing enough food for the camp was a local challenge. Hal B. Fullerton organized the Market Committee of the Long Island Food Reserve Battalion and met with farmers to discuss the best methods of supplying food for the camp. Civilian laborers are seen here unloading flour for the camp bakery. They are stacking the sacks within a tent frame in the supply area under the watchful eyes of soldiers.

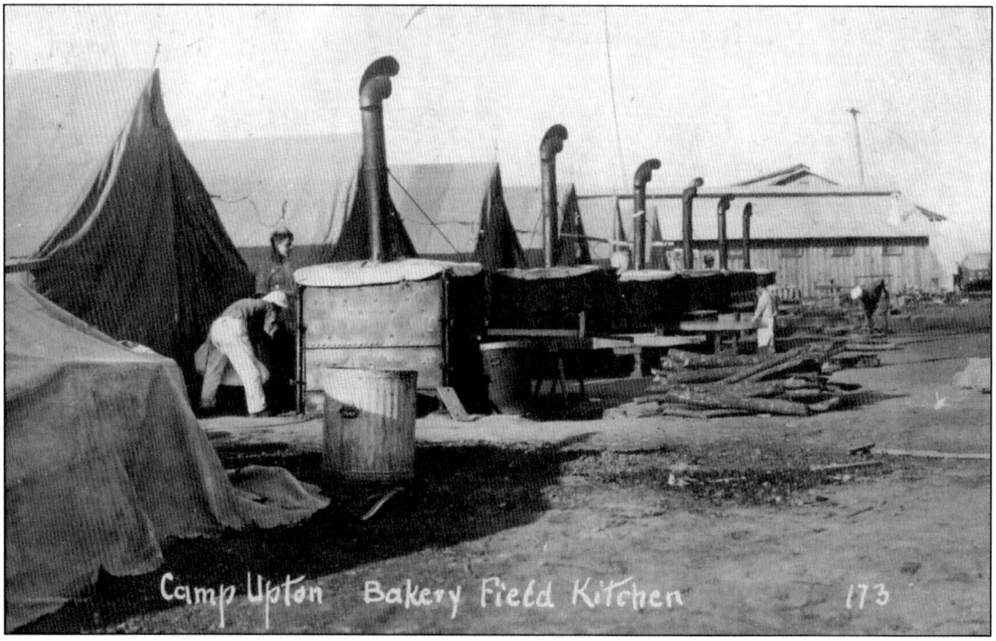

Camp Upton Bakery Field Kitchen 173

Referred to as "the real doughboys," these men are baking bread in the field kitchen bakery to distribute to the various barracks. When the camp bakery was constructed, it produced 25,000 pounds of bread a day.

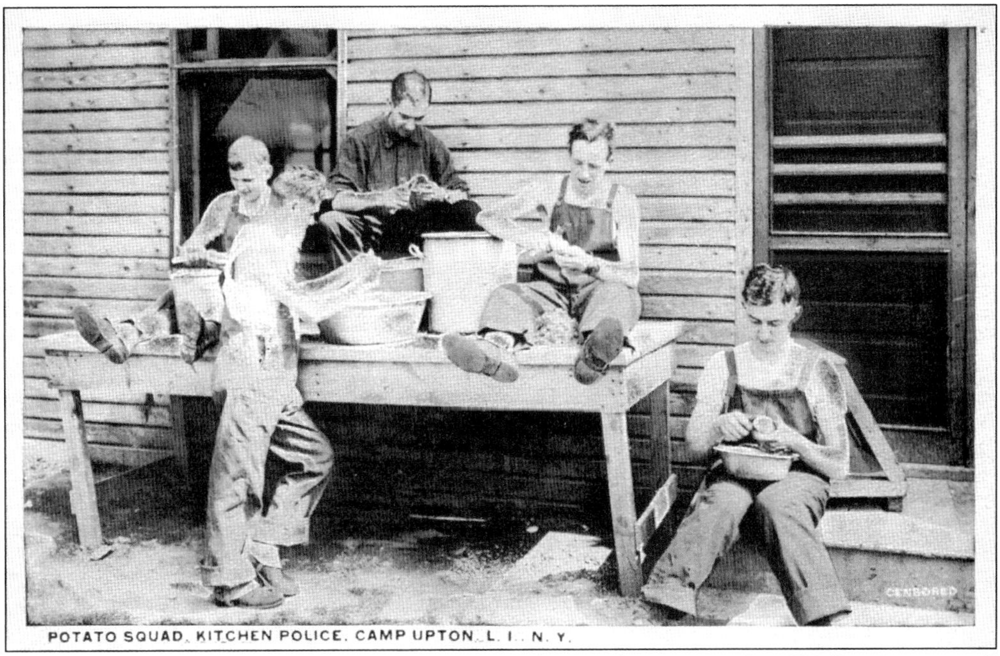

Peeling potatoes was a daily duty for many enlisted men when assigned to KP, "kitchen patrol" or "kitchen police," temporary work under the direction of the kitchen staff. KP duty could also include mopping and sweeping floors, clearing tables, washing dishes, and helping prepare the next meal.

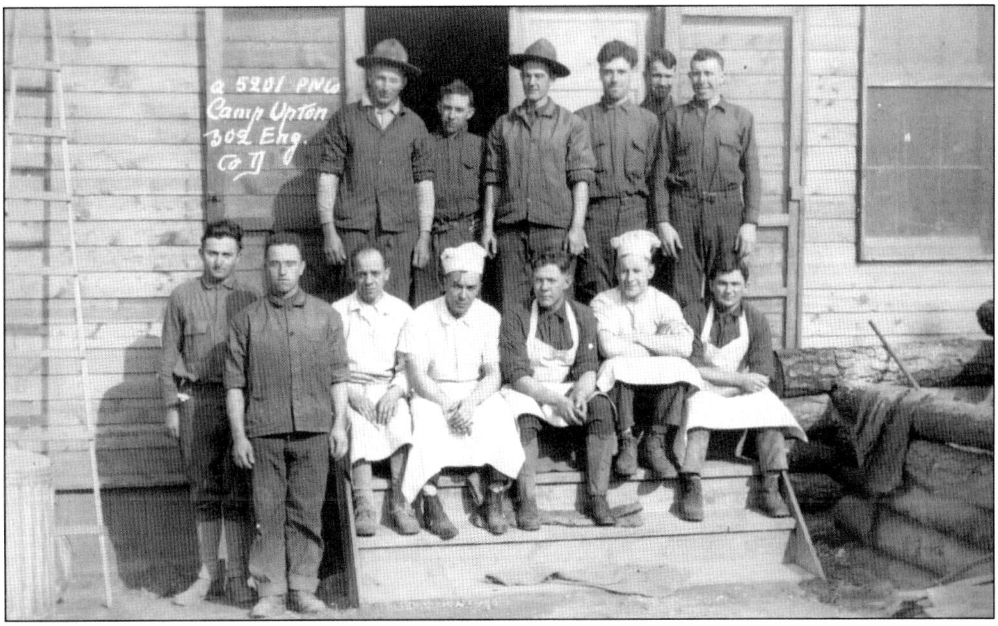

Cooks (in white), aided by the KP men, needed to prepare food for a wide variety of tastes due to the diverse mix of soldiers at the camp, which was nicknamed "Camp Melting Pot." One soldier wrote, "Dear grandmother: you would laugh to see the stuff that is turned out of this place in a day. . . . I just had macaronia with tomato, bread pudding, bread and butter and cup of tea, some feed. It sure was good. Harry."

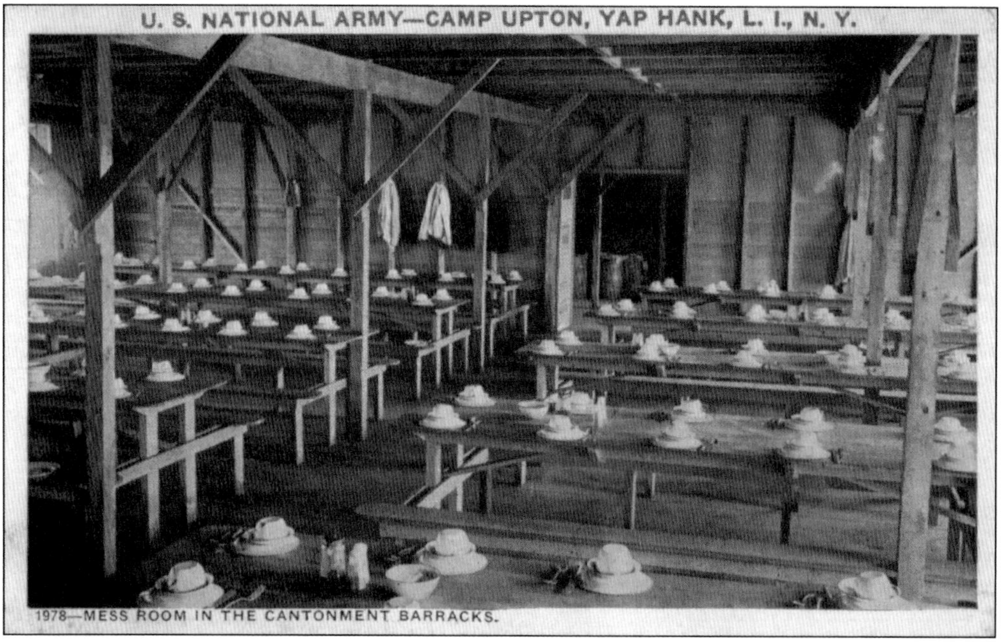

Each barracks building had its own mess hall with cooks who prepared and served the meals on the first floor, while soldiers bunked on the second floor. The first floor was heated with stoves; there was no heat on the second floor.

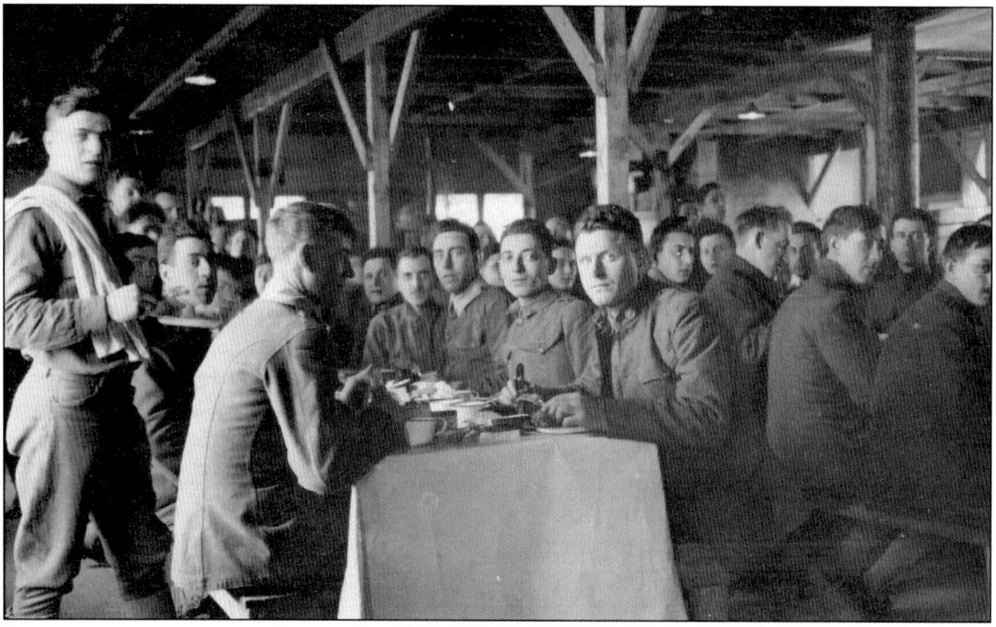

The mess hall was located on the first floor of each barrack. Enlisted men are shown here eating a meal at mess tables in the dining area on the first floor of a Company Barracks building. The second floor of each barracks was used for sleeping quarters.

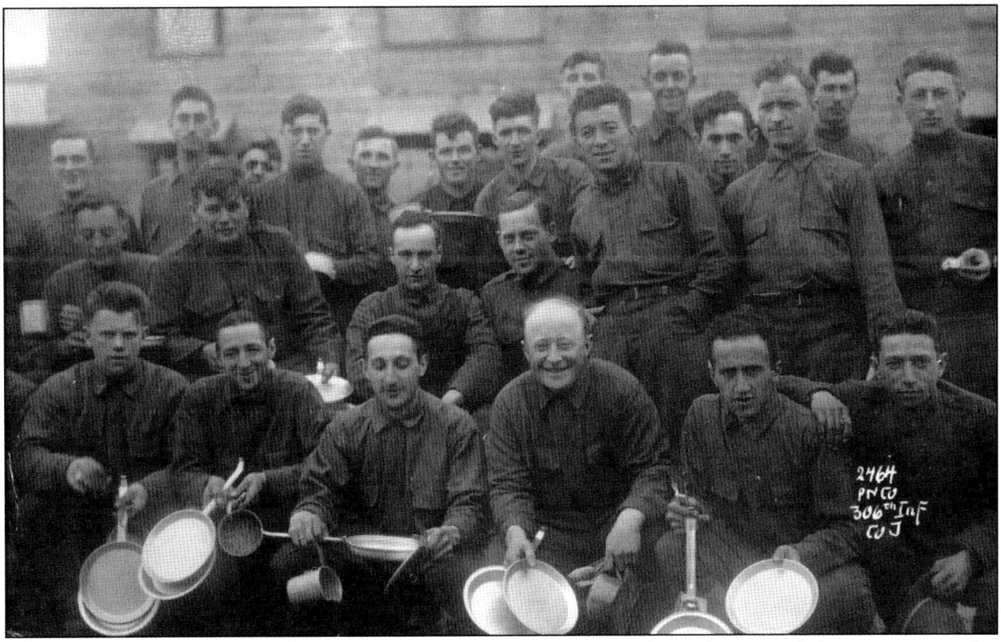

One of the first things given to recruits was a two part mess kit, a knife, a fork, a spoon, and a cup. They used these utensils at all times to eat their meals. These men are washing their mess kits after a meal.

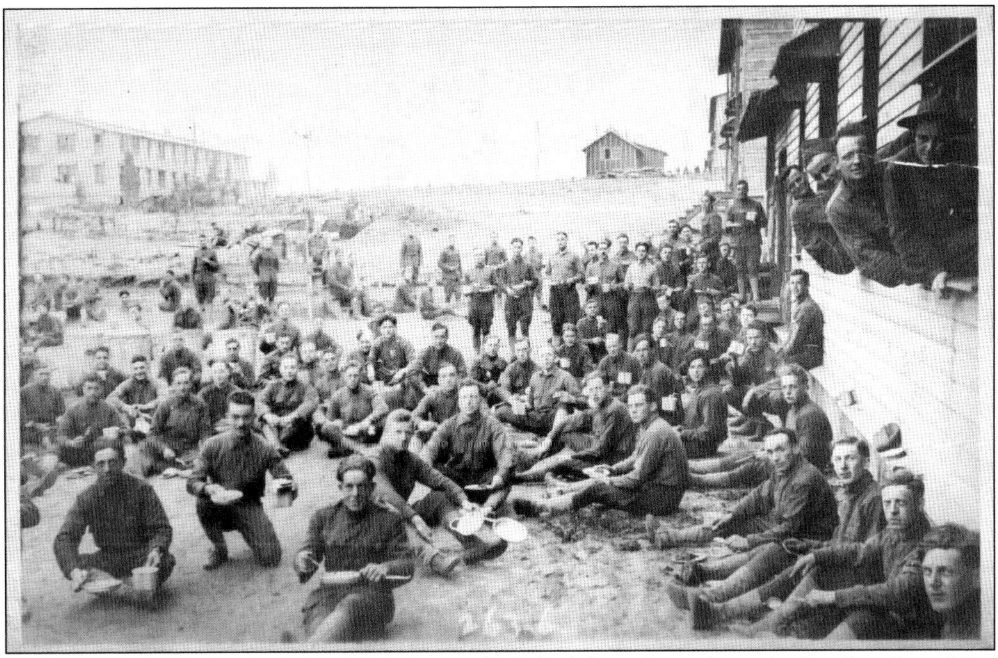

The first meals at Camp Upton were prepared by chefs brought in from the best hotels in New York City. The officers even preferred to eat with their men instead of at their own barracks.

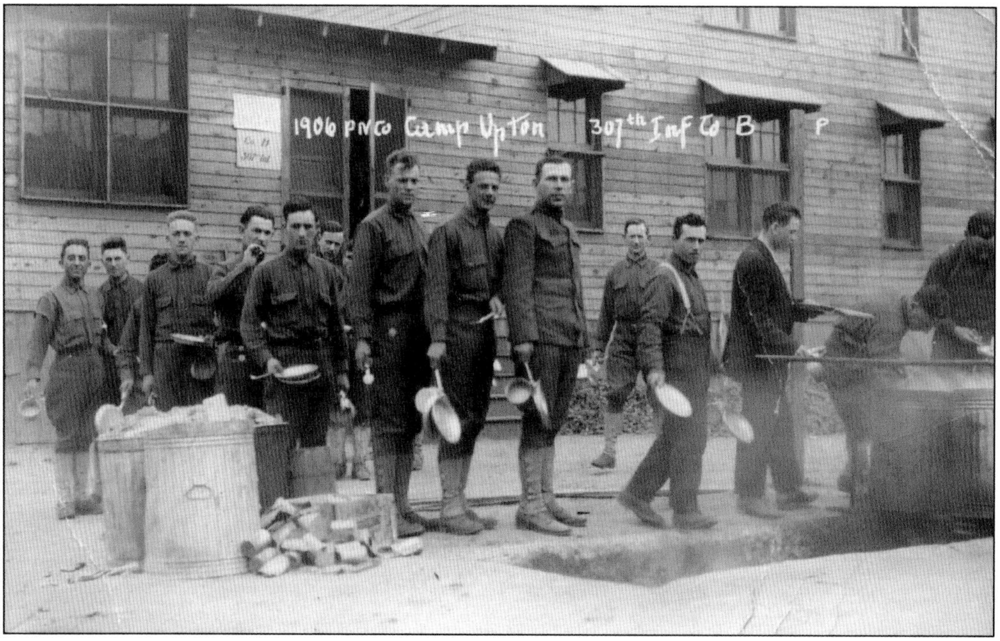

Soldiers line up to wash out their mess kits in garbage cans of boiling water and soap, over an open fire in a pit. Cleanliness was paramount in such close quarters. Garbage was burned at the camp after local farmers could no longer handle the volume that needed to be hauled away.

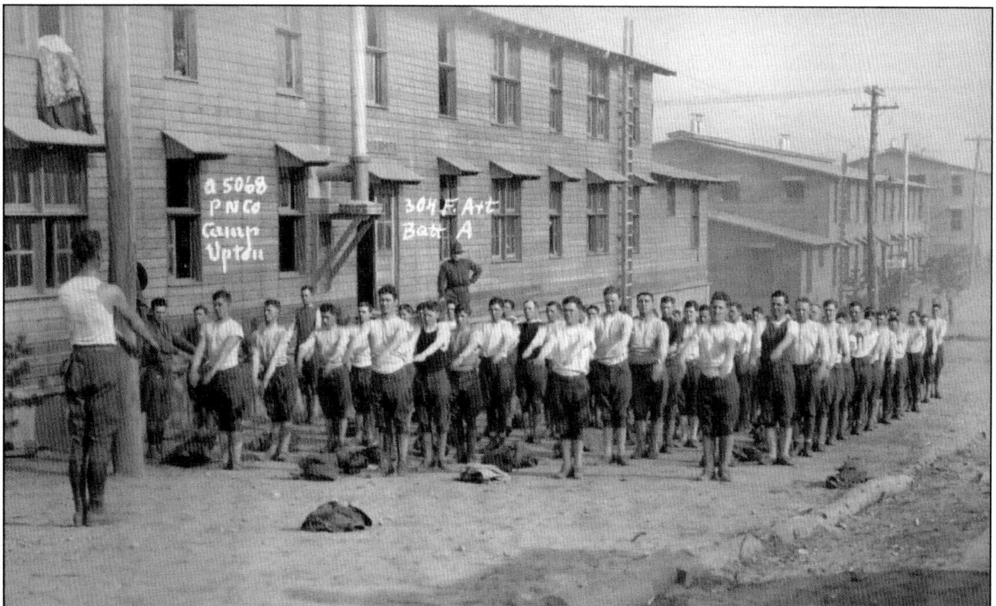

This photograph shows enlisted men of Battery A of the 304th Field Artillery exercising outside their barracks in the late summer of 1917. This was an important activity which improved unit cohesion, reinforced discipline and improved the physical condition of the soldiers.

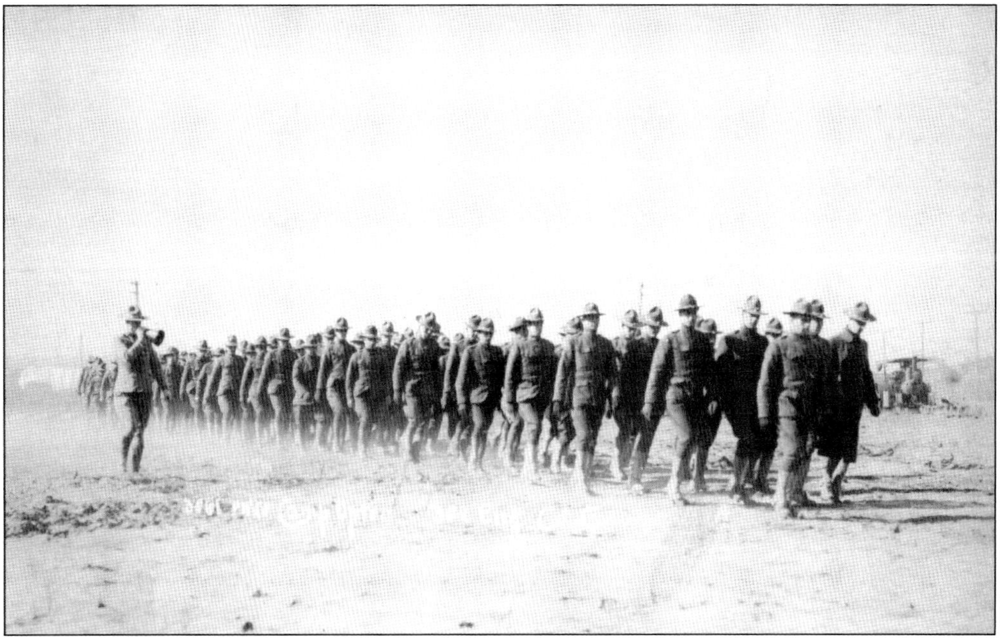

The 302nd Engineers, the engineer regiment of the 77th Division, are shown here on the march. This photograph was taken soon after the recruits arrived. The officer is counting cadence to keep them in step and a construction tractor is visible in the background on the right.

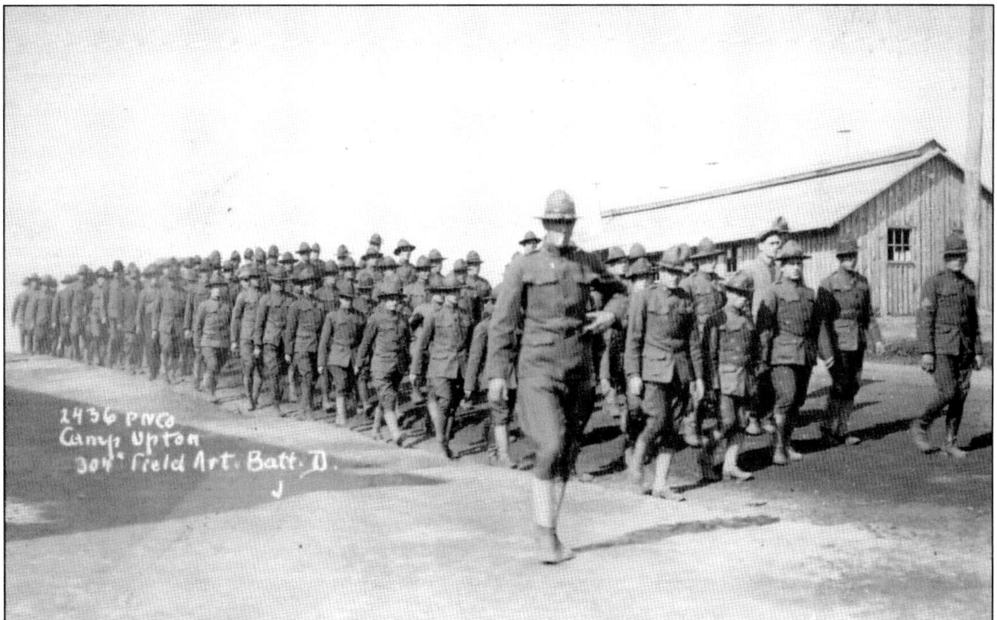

Battery D, 304th Field Artillery set off on a hike. Hiking was another activity that was a part of beginning military training. It improved physical condition and enforced discipline without requiring military equipment.

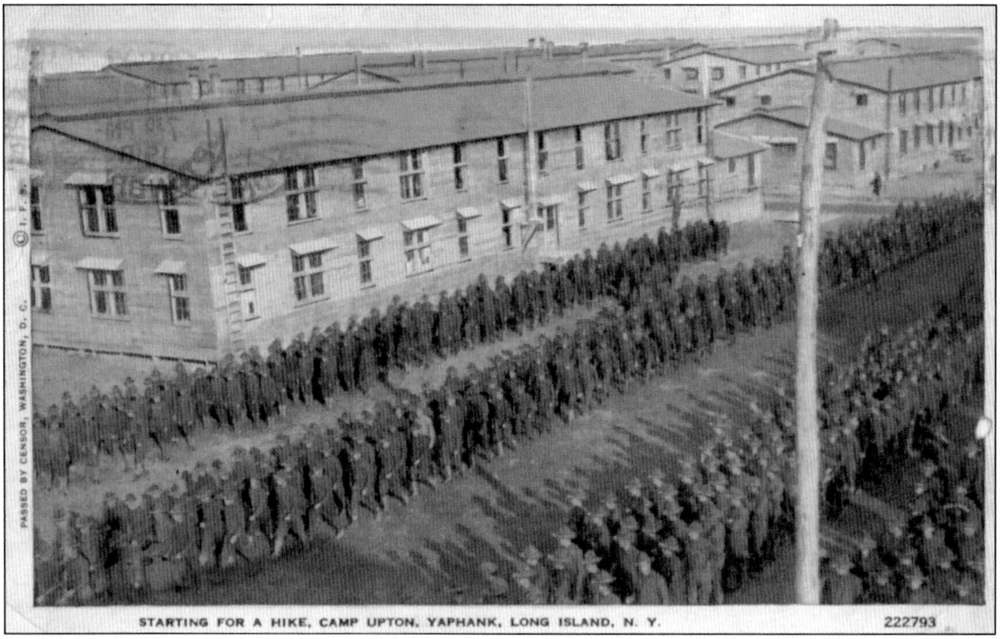

After a short time, the recruits had become accustomed to marching and hiking in small units. Hiking units were then enlarged to the size of battalions, which were made up of four companies and the three battalions of a full regiment.

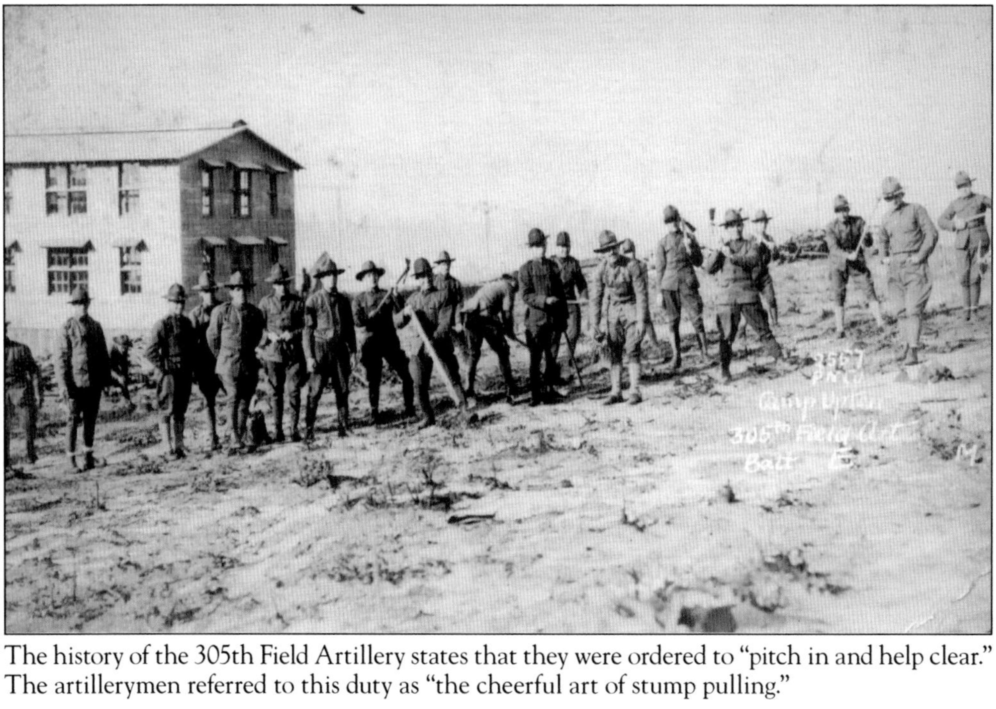

The history of the 305th Field Artillery states that they were ordered to "pitch in and help clear." The artillerymen referred to this duty as "the cheerful art of stump pulling."

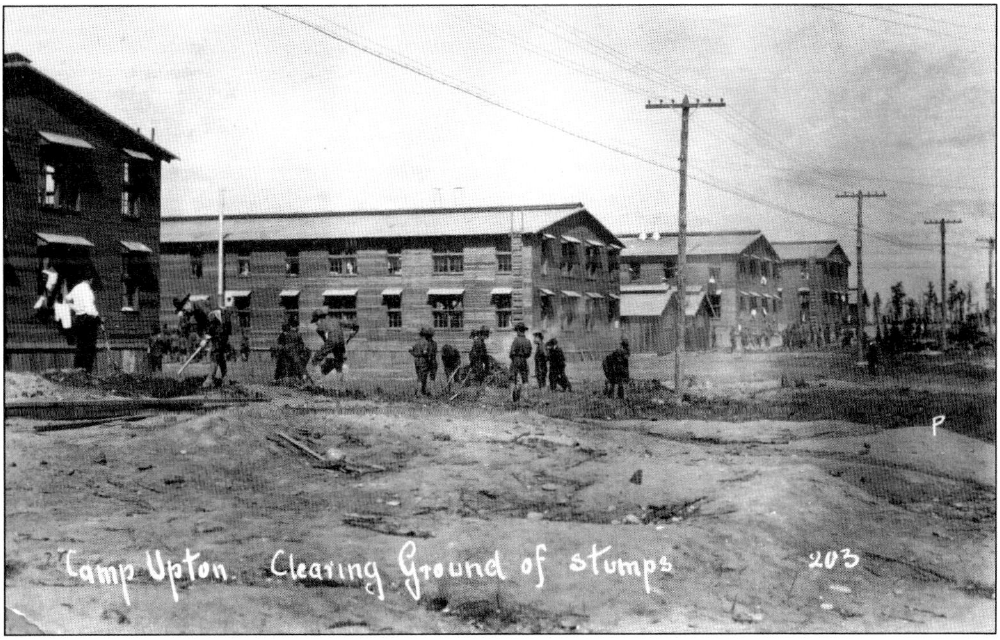

Construction quartermaster Gen. O'K. Myers reported on construction of the camp in 1918 that soldiers encountered "stumps six feet in diameter," remains of the old-growth forest that had been cleared for cordwood in the 19th century. What they found was scrub oak, pines, and tangled undergrowth.

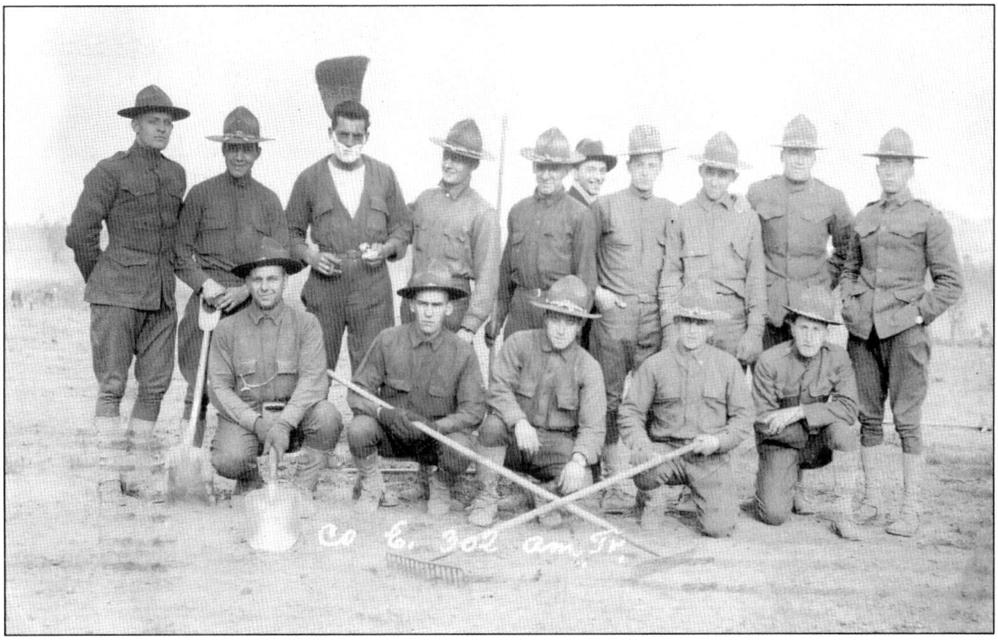

In addition to their military training schedule and working to complete clearing of the cantonment grounds, recruits found time for some joking around. Perhaps the crossed rakes indicate a new branch of the service.

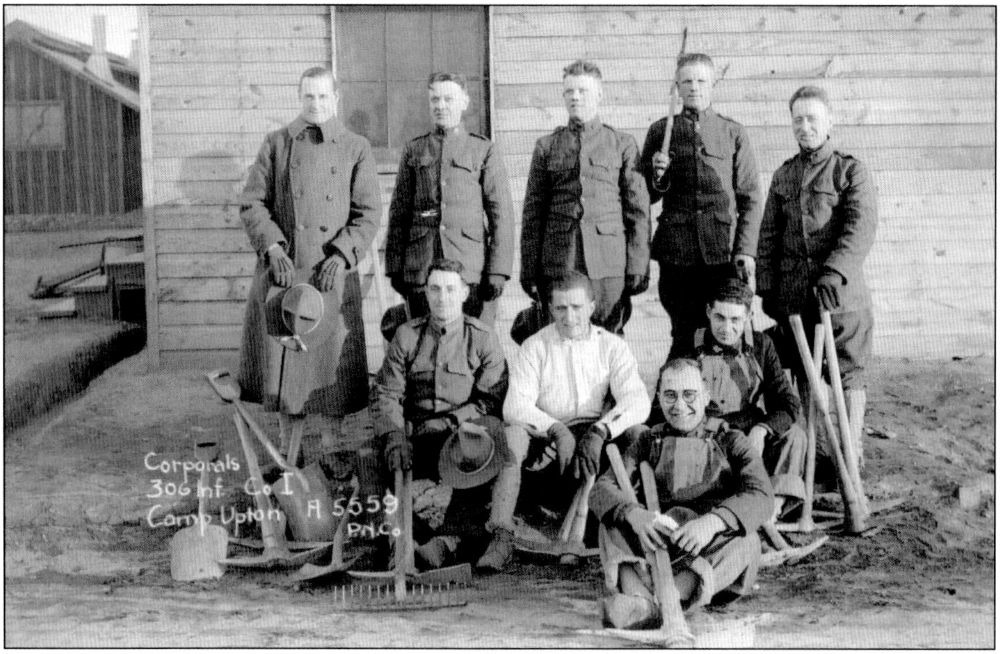

A group of corporals stands with the weapons used to conduct what the recruits referred to as their first offensive on the "No Man's Land of brush and trees." Foresters working on the construction contract cut down 1,000 trees a day, leaving the stumps.

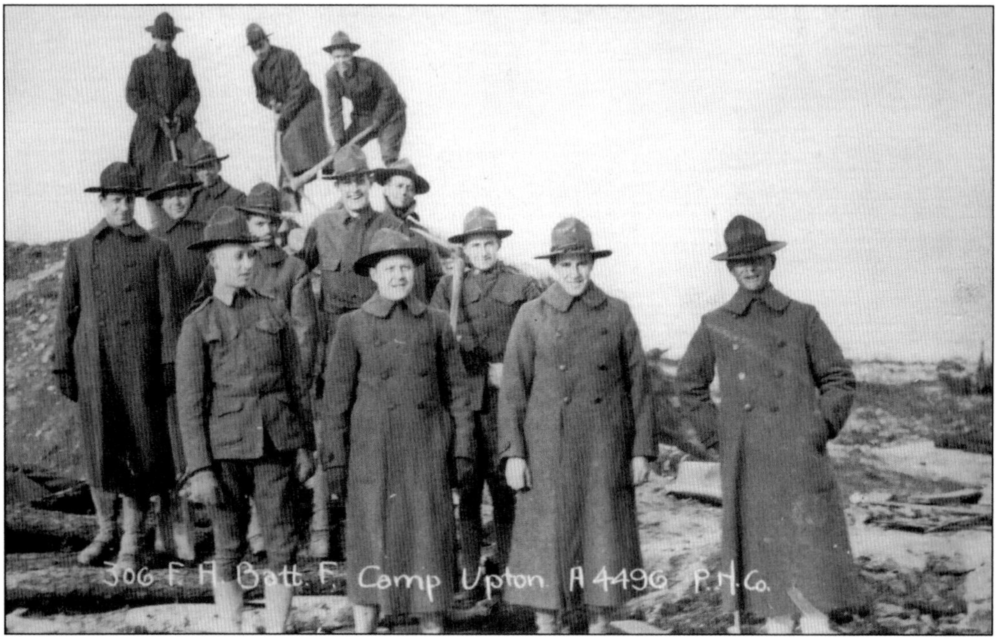

Members of the 306th Artillery are carrying pickaxes and shovels and are dressed in a mix of overcoats and tunics. This photograph and others show that the clearing of stumps continued and the task of moving coal was added during the fall and winter of 1917.

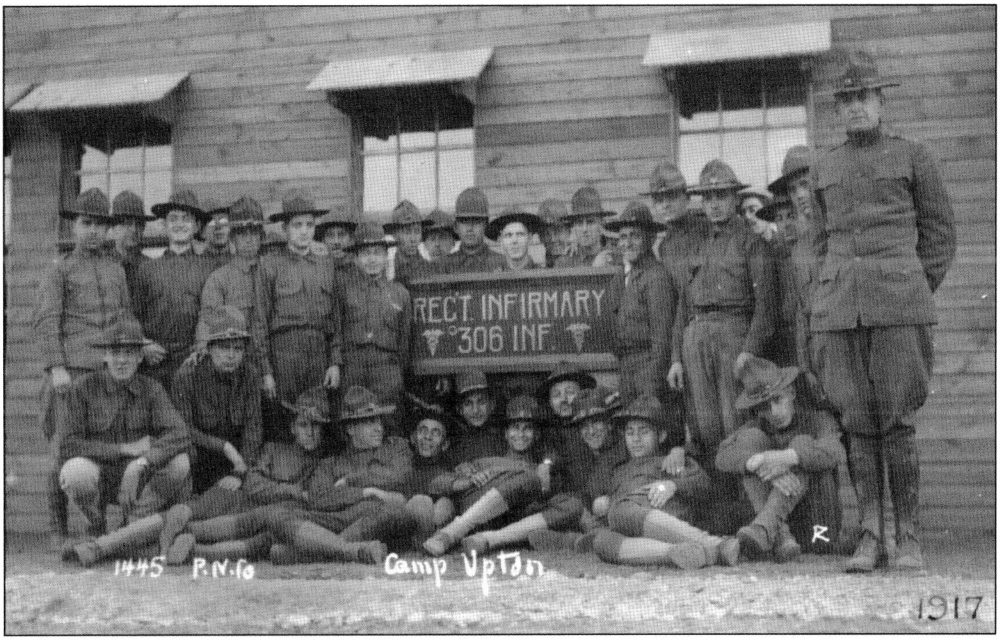

Each regiment had its own infirmary. Soldiers who felt ill at "sick call" were marched to the infirmary, where they would be examined, and the doctor would determine whether they were to be cared for there or at the base hospital.

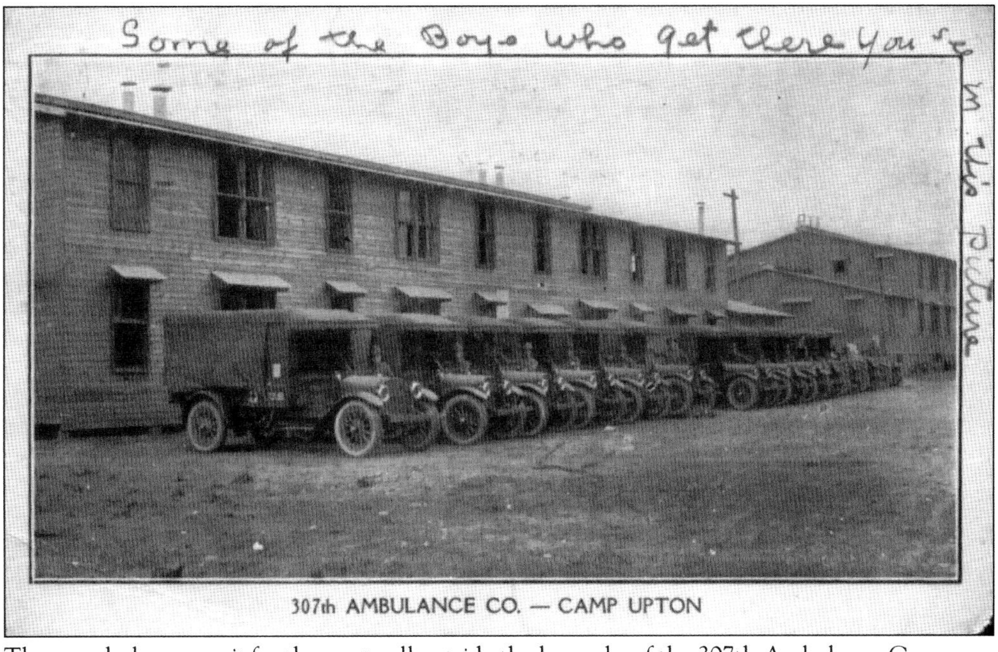

These ambulances wait for the next call outside the barracks of the 307th Ambulance Company. Each infantry regiment had its own ambulance company. The drivers are enlisted men of the Medical Corps.

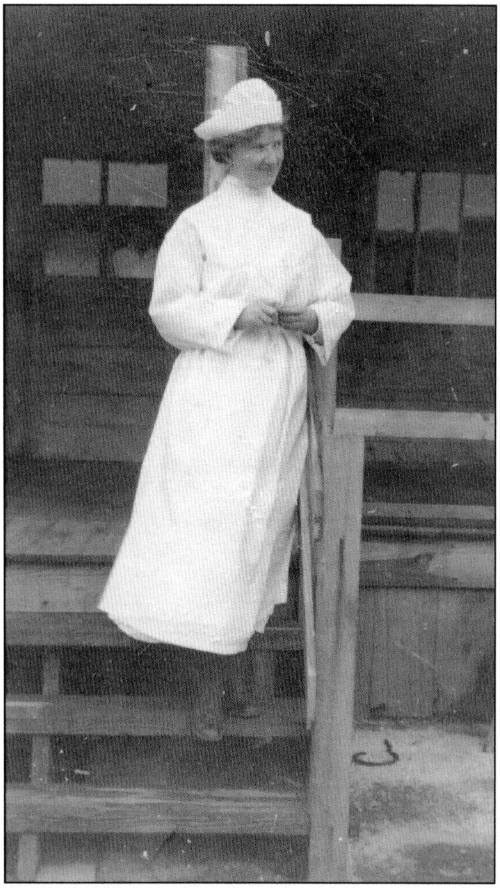

This photograph depicts Mary Cecilia Kennedy from Poughkeepsie, a nurse at Camp Upton. She was born on May 4, 1894, and died in 1978. She was one of 350 nurses employed by the base hospital. The Red Cross staffed hospitals and ambulance companies and recruited 20,000 registered nurses to serve the military.

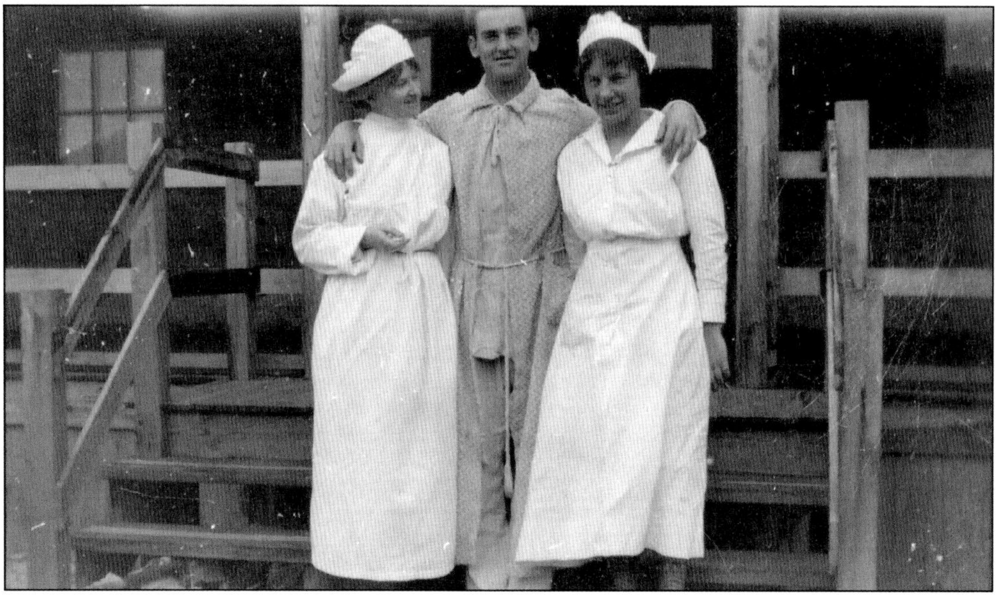

Kennedy indicated she was in this photograph (on left) with "patient with mumps and Miss Haimly." Measles was another infectious disease that soldiers were vaccinated for.

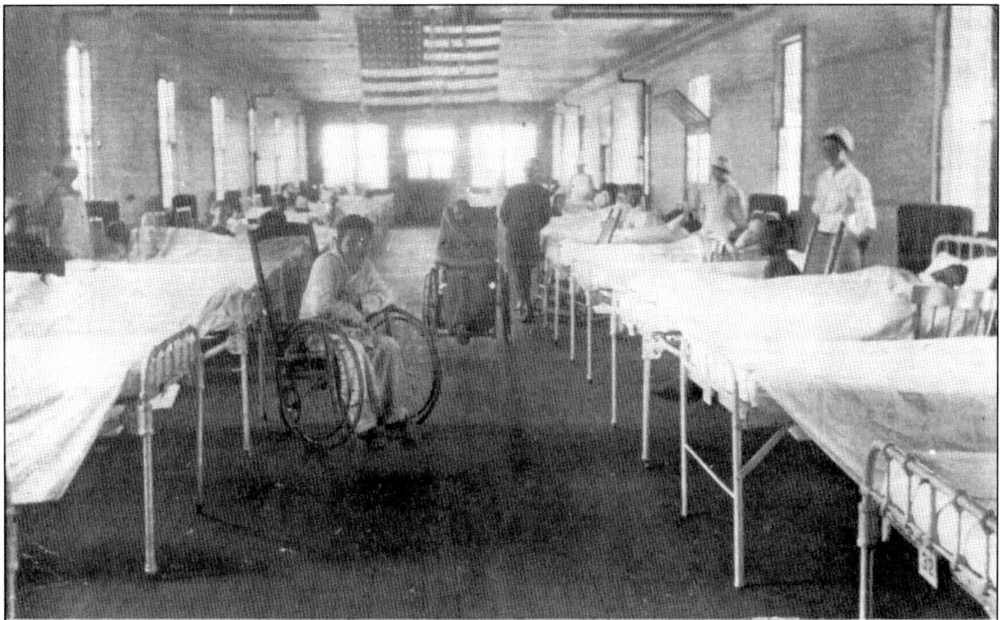

A TYPICAL WARD, U. S. A. Base Hospital, Camp Upton, L. I.

The base hospital was located in a separate area about three miles from the camp proper. It consisted of 115 buildings with a staff of 1,500, including 140 doctors and 350 nurses. By the end of the war, over 360,000 cases had been treated.

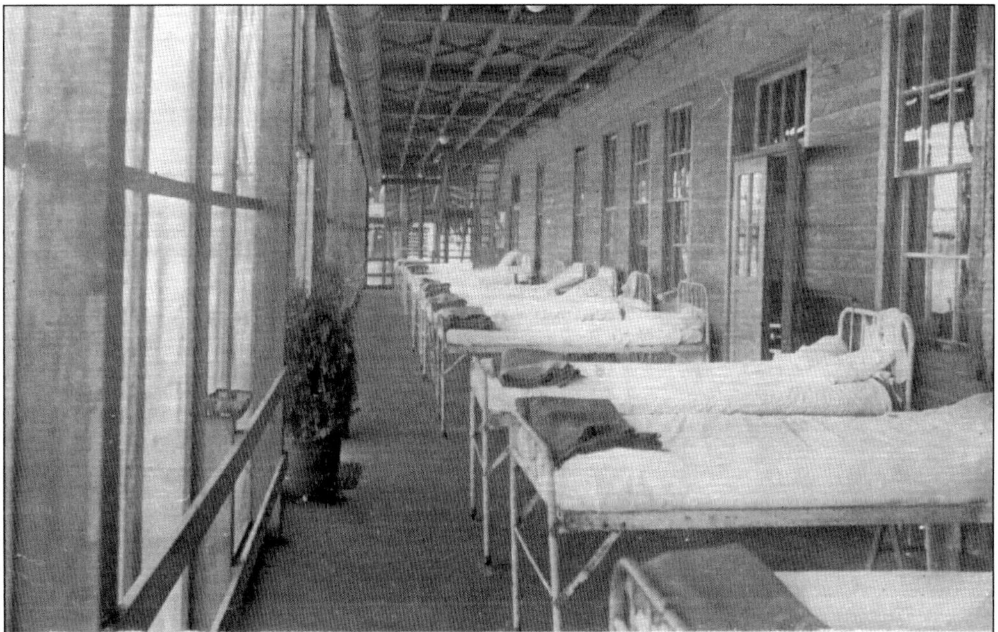

SLEEPING PORCH, U. S. A. Base Hospital, Camp Upton, L. I.

The buildings of the hospital were connected by glass-enclosed passageways. The photographer shot through the screen for this image. The screen protected the patients from the serious problem of mosquitoes in the summer at Camp Upton.

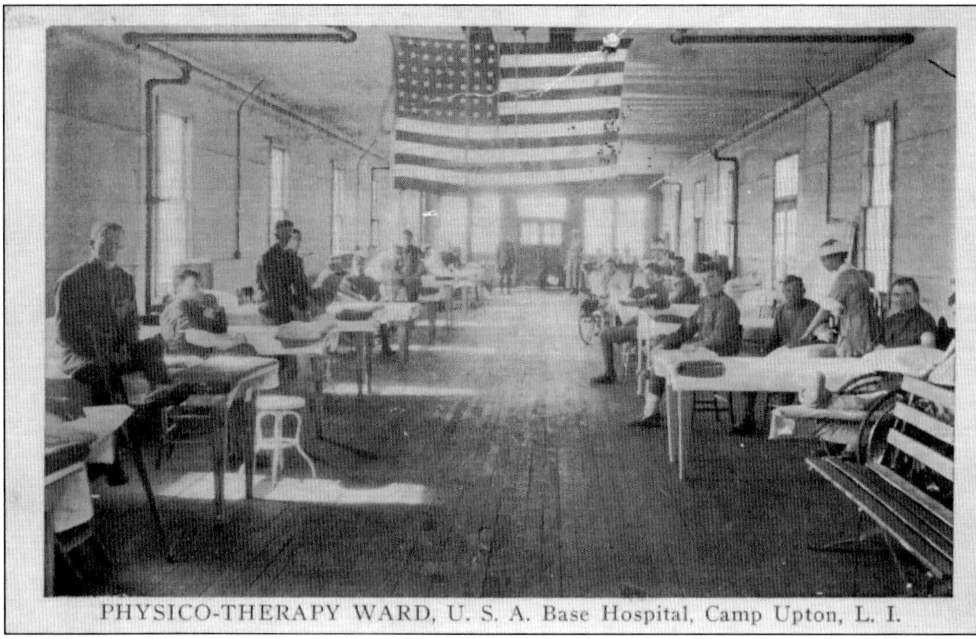

The influenza pandemic struck Camp Upton in September 1918, affecting more than 6,000 soldiers. Many then developed pneumonia, resulting in over 400 deaths before ending its course in late October. At first, burials took place in a cemetery at Camp Upton.

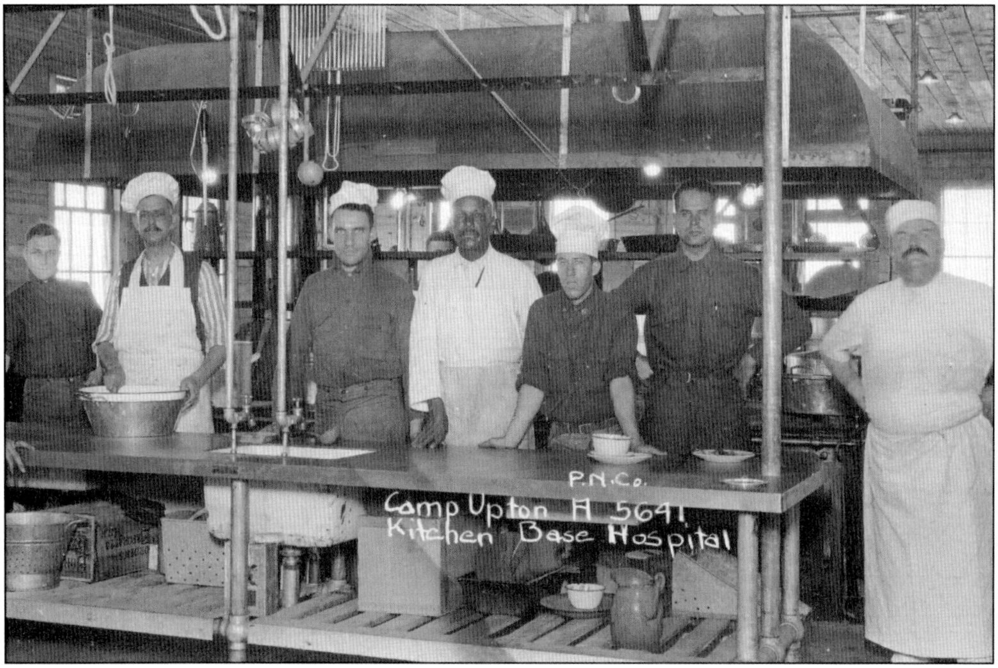

Kitchen crews at the base hospital needed to prepare meals for many different diets, including kosher food for Orthodox Jews. There were more Jewish soldiers in Camp Upton than at any other camp in the country.

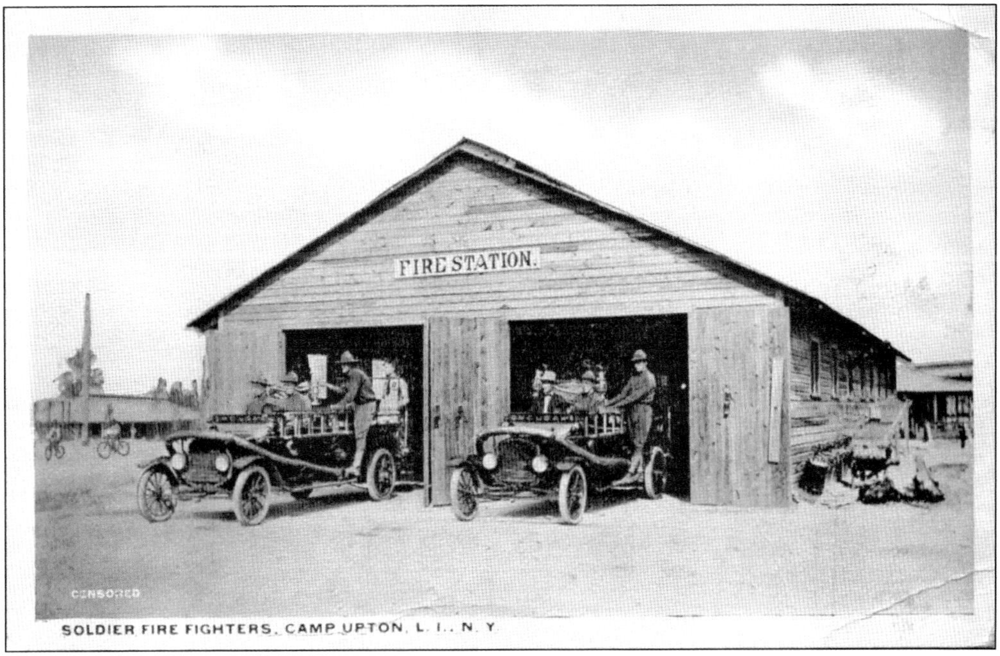

The fire department was outfitted with two American LaFrance triple combination pumping engines, two Brockway combination hose wagons and chemical wagons, a Seagraves combination pumping engine, a Dodge chemical wagon, and four Howe-Ford combination hose wagons and pumps.

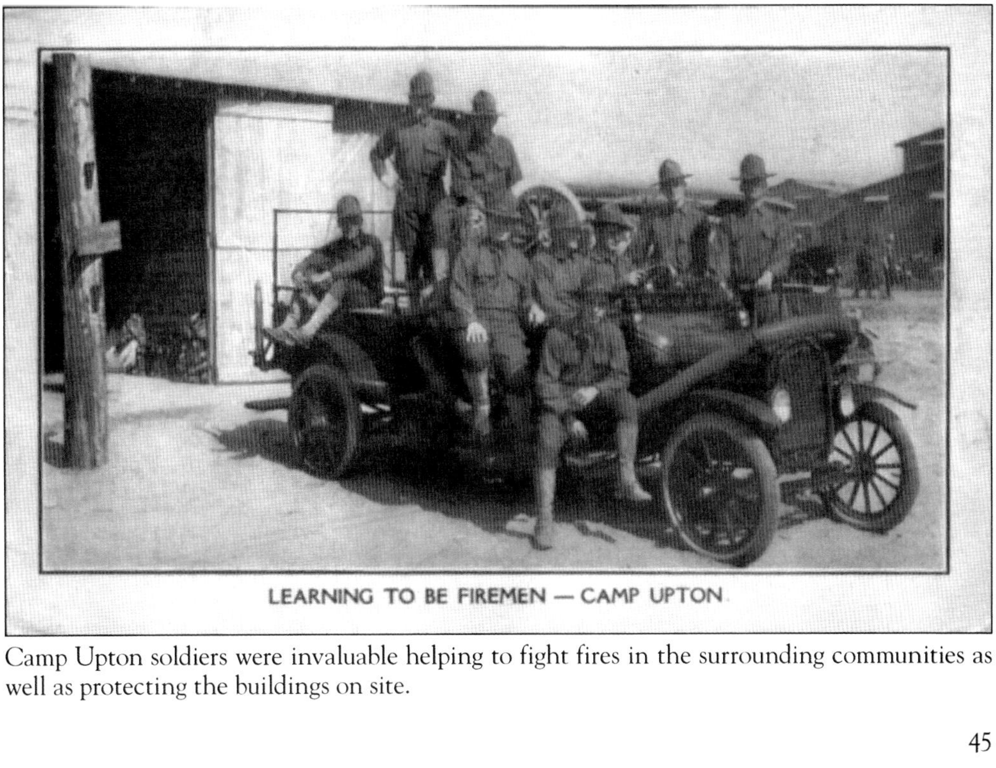

Camp Upton soldiers were invaluable helping to fight fires in the surrounding communities as well as protecting the buildings on site.

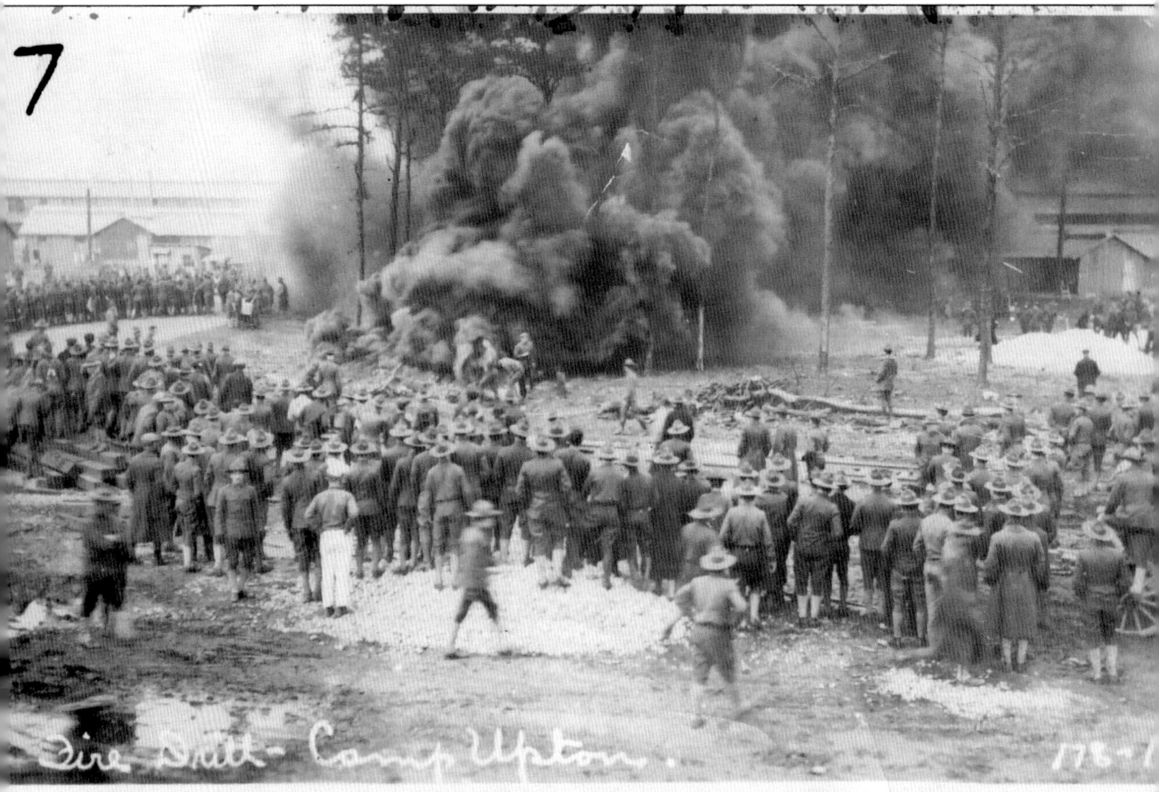

A fire at camp was a constant danger, as all of the buildings were wooden. More than 45 New York City firemen served as soldier firefighters at the camp, forming three double companies. The camp was divided into fire districts and had numbered fire alarm boxes.

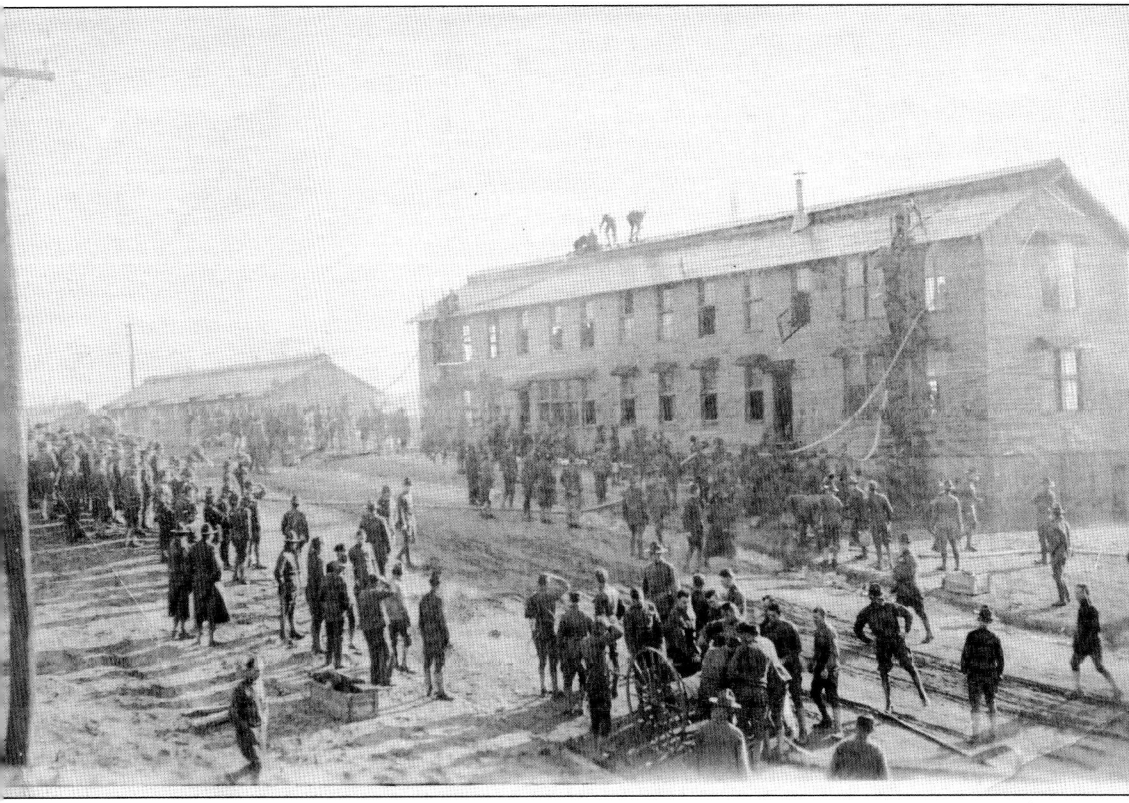

Pictured here is a fire drill involving a barracks building. These were very important, as the barracks were heated with pot-bellied stoves filled with coal and the men slept on straw mattresses in a building constructed of pine lumber. Fire drills were essential.

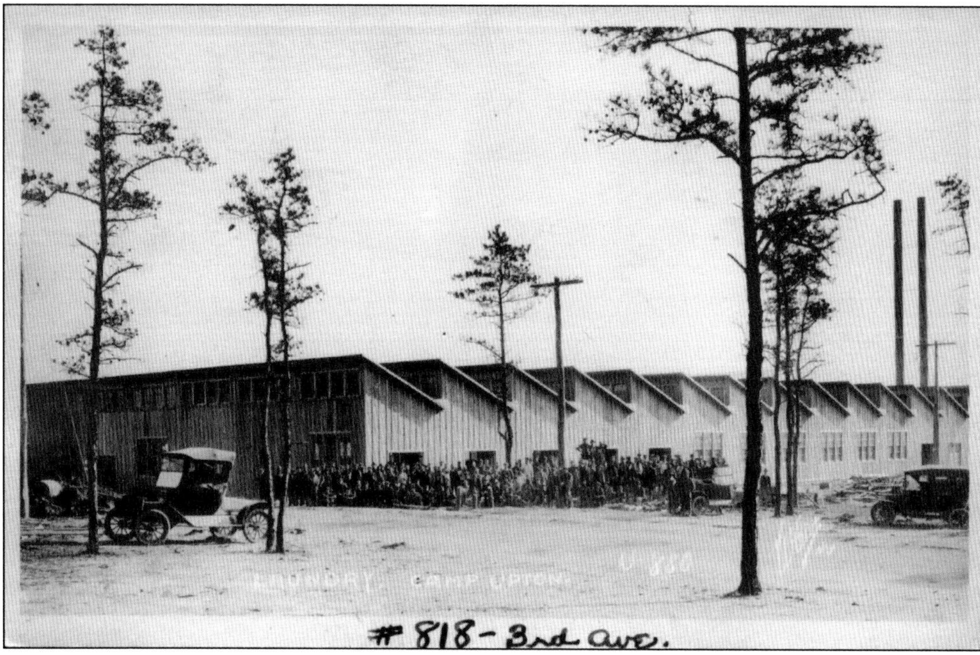

The cantonment laundry building at Camp Upton claimed to be "the largest in the world." Six hundred civilians did the laundry of 50,000 men. Over 1.5 million items were washed each month. Clothing and bedding was also sent from Camp Mills in Garden City before it built its own facility.

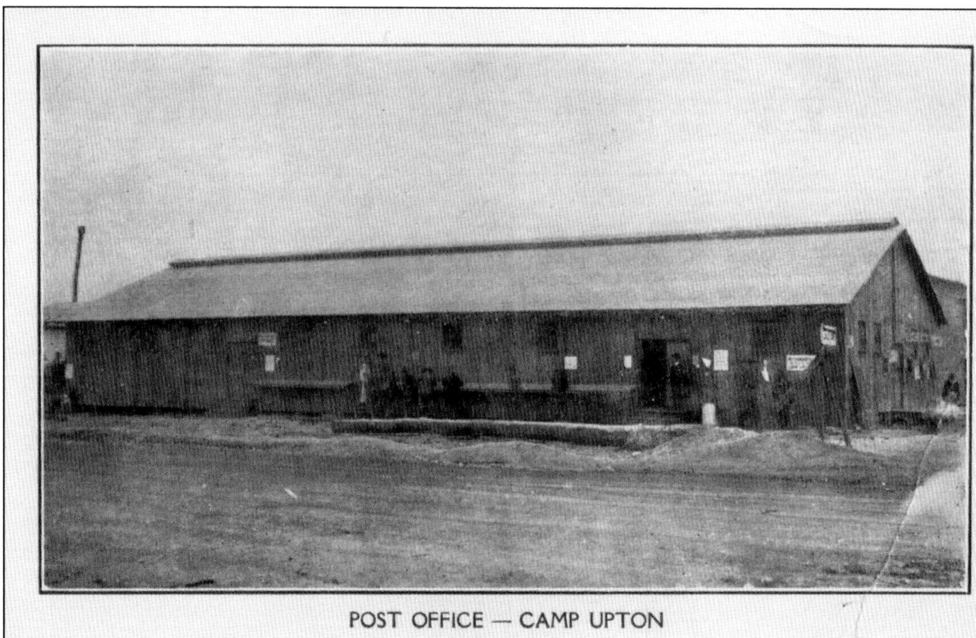

The Upton branch of the Brooklyn Post Office was established on July 21, 1917, and was discontinued on July 31, 1920. The post office operated with 60 clerks and 40 soldiers at five service windows. At its height, 50,000 pieces of mail a day were handled. Letters and postcards were marked "Mail leaving Camp Upton."

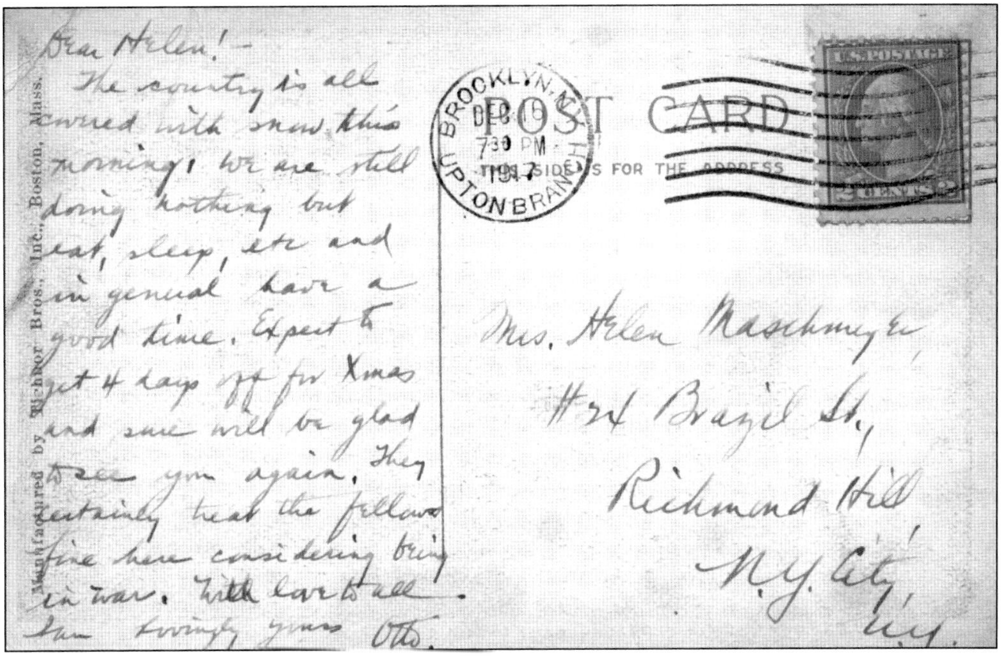

Dear Helen! -
The country is all covered with snow this morning. We are still doing nothing but eat, sleep, etc and in general have a good time. Expect to get 4 days off for Xmas and sure will be glad to see you again. They certainly treat the fellows fine here considering being in war. With love to all
I am lovingly yours Otto.

BROOKLYN, N.Y. DEC 16 7:30 PM 1917 UPTON BRANCH

POST CARD

THIS SIDE IS FOR THE ADDRESS

U.S. POSTAGE 2 CENTS

Mrs. Helen Maschmeyer,
#24 Brazil St.,
Richmond Hill
N.Y. City,
N.Y.

Otto Maschmeyer sent this postcard on December 16, 1917. The men were given a half-holiday between Christmas and New Year's. Half the division could go home, and when they returned, the other half went out. The cooks reported they only had enough food to feed half the camp at once over the holidays.

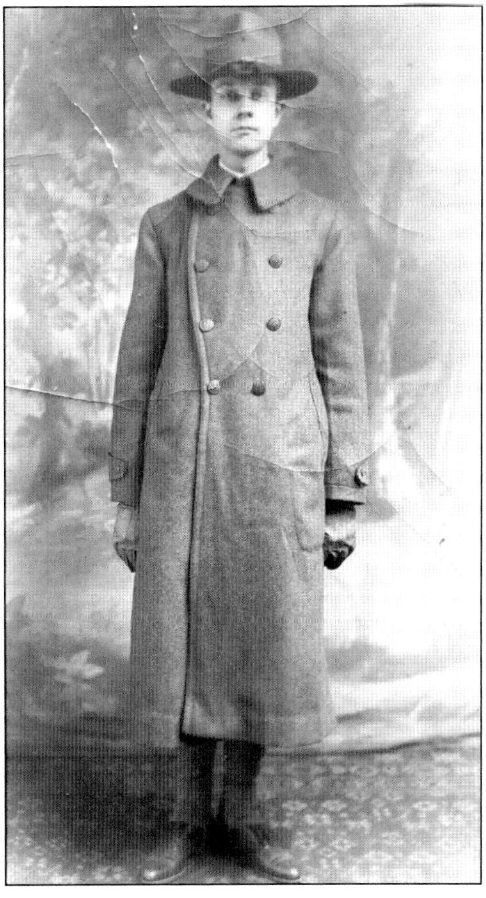

This soldier named Henry, dressed in his heavy Army overcoat, sent his photograph with New Year's greetings to his family from the camp. Once he reached France, his beloved campaign hat would be replaced by a steel helmet.

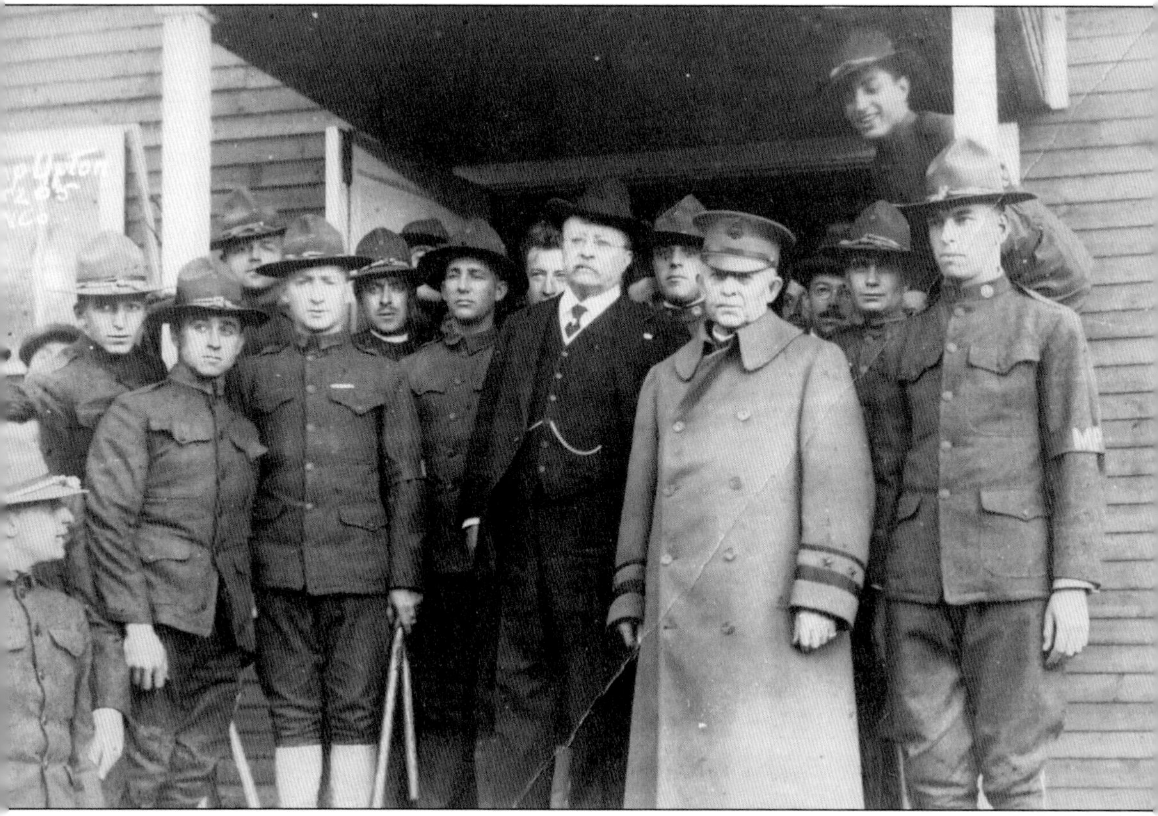

By October 1917, 30,000 men had arrived at the camp, doubling the population of Suffolk County. New York sent more soldiers to fight than any other state in the Union. Camp Upton was officially opened on December 20, 1917. Former president Theodore Roosevelt visited Camp Upton on November 18, 1917, just before its official opening. The president gave two speeches: one to 3,000 soldiers at the YMCA auditorium and another to the black soldiers at the Knights of Columbus hall. Roosevelt fiercely opposed Pres. Woodrow Wilson's neutrality policy and urged military preparedness. To the right of Roosevelt is Camp Upton's commanding officer, Maj. Gen. J. Franklin Bell, with whom Roosevelt had fought in the Spanish-American War. President Roosevelt died at Sagamore Hill in Oyster Bay in 1919.

Two

This Is the Army

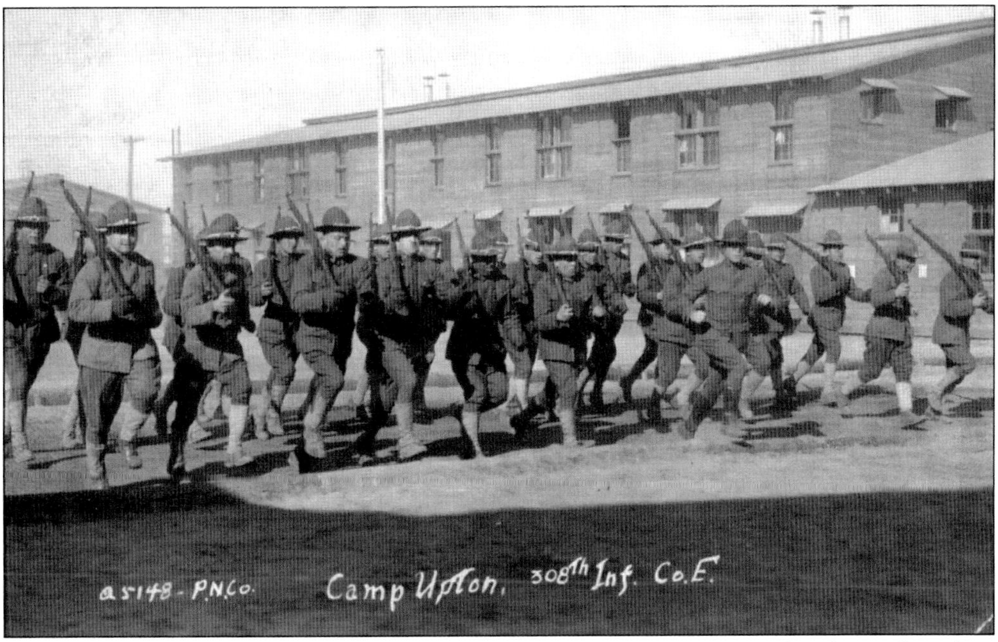

Soldiers spent 16 weeks in intense training before departing for Europe. They learned modern trench warfare techniques using bayonets, hand grenades, machine guns, and hand-to-hand combat, as well as how to shoot a rifle. American officers insisted that enlisted men learn mortars, cannons, and rifle marksmanship and small-unit tactics.

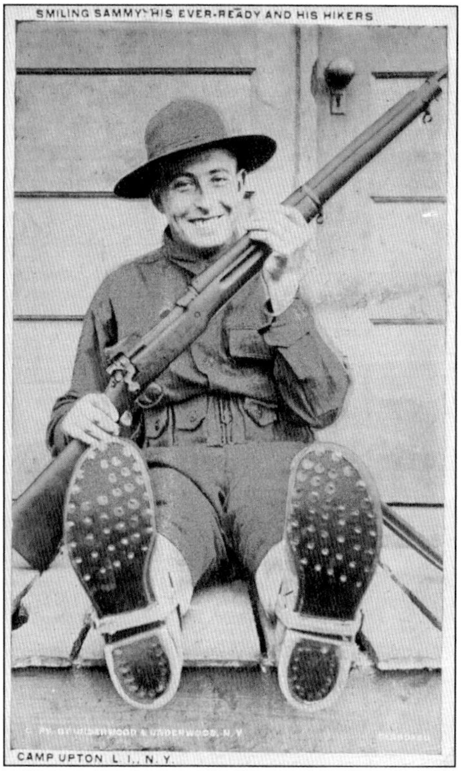

Here is "Smiling Sammy" in his newly issued uniform holding his US Model 1917 rifle and his Mills cartridge belt. The soles of his boots are covered with hobnails to provide traction and to extend their usable life.

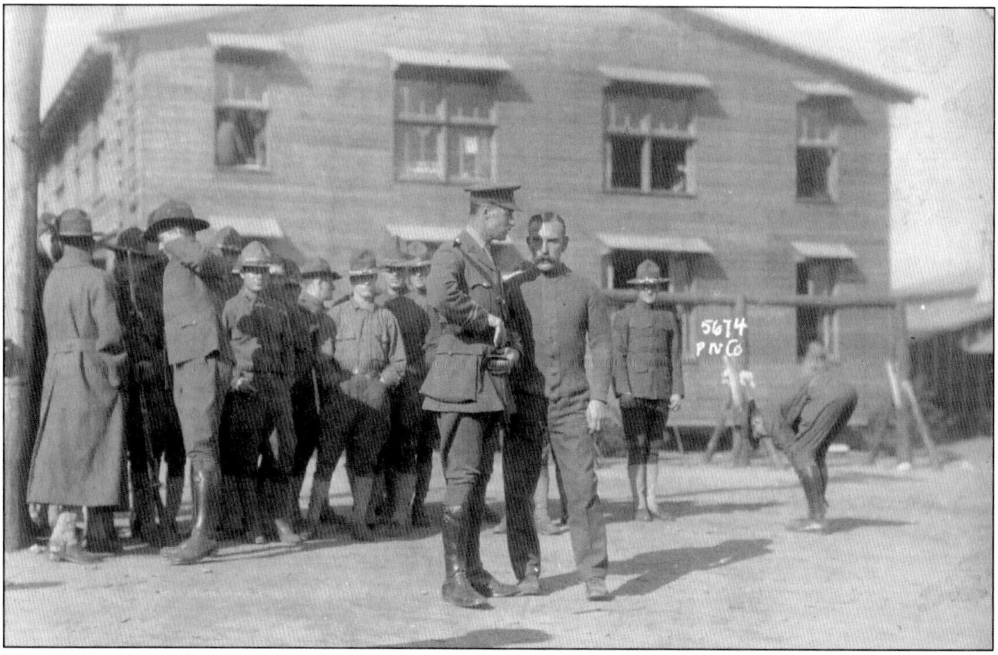

British and French officers were brought to US training camps to help train the officers and enlisted men. They demonstrated weapons, helped with drill, and presented lectures to officers in the evenings. This photograph shows Captain Browne and Sgt. Maj. G.C. Covington of the British Army.

Pictured here is Cpl. Harold C. Steiger, 302nd Ammunition Train, Company G. Corporal Steiger was from Saugerties, New York. He is in light marching order with his rifle with bayonet attached, Mills cartridge belt, and first aid kit.

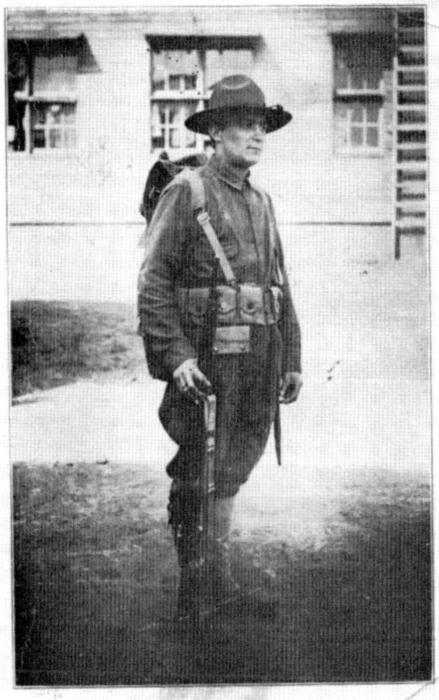

Corporal HAROLD C. STEIGER, 302 Ammunition Train
Camp Upton, Long Island

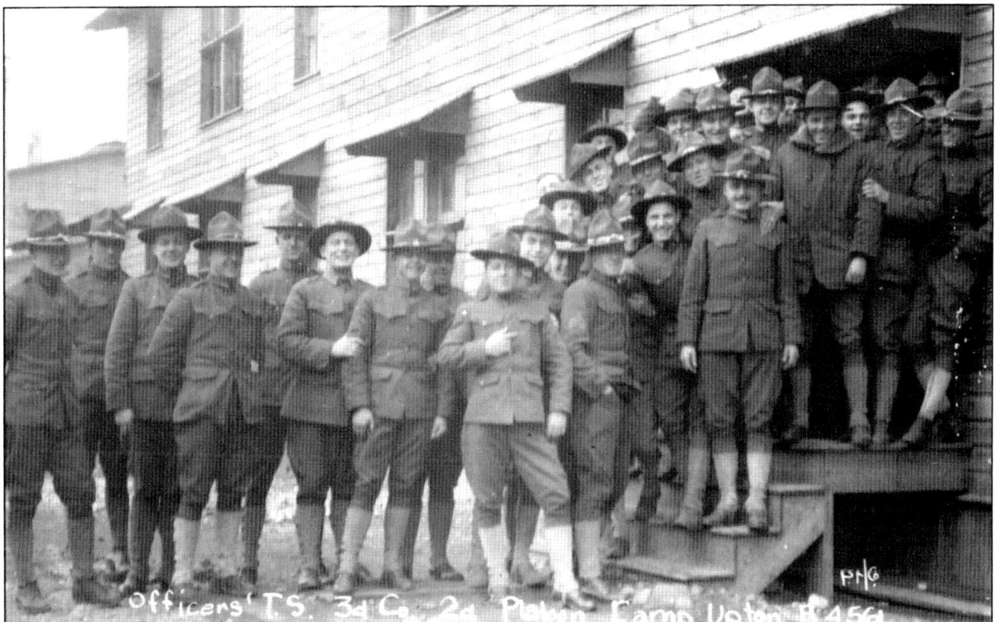

Military units of the new 77th Division required officers, and the training camp at Plattsburgh did not train a sufficient number to meet the need. In order to solve this problem, an officer training school was established at Camp Upton in January 1918.

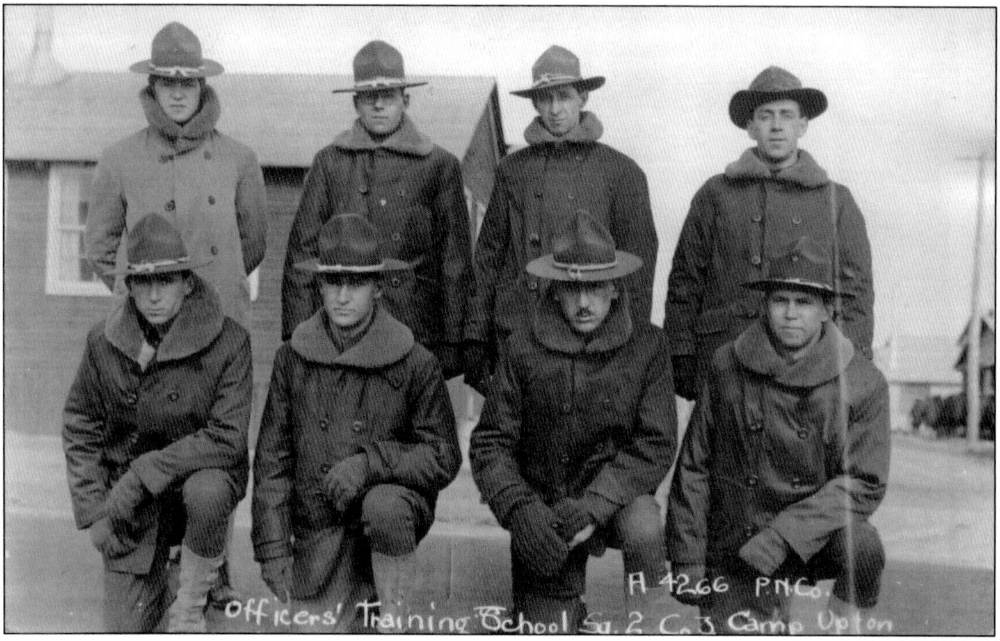

Promising noncommissioned officers (sergeants and corporals) were assigned to squads of eight men, which in turn formed platoons and companies. Graduates of the school could be commissioned as officers in the US Army.

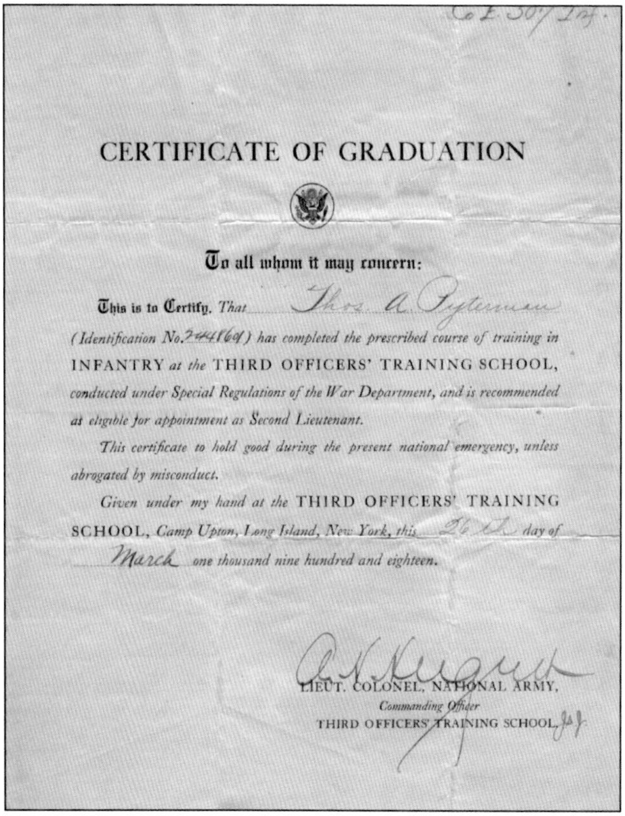

Co E 307 Inf.

CERTIFICATE OF GRADUATION

To all whom it may concern:

This is to Certify, *That* Thos. A. Pyterman *(Identification No.* [illegible]*) has completed the prescribed course of training in* INFANTRY *at the* THIRD OFFICERS' TRAINING SCHOOL, *conducted under Special Regulations of the War Department, and is recommended as eligible for appointment as Second Lieutenant.*

This certificate to hold good during the present national emergency, unless abrogated by misconduct.

Given under my hand at the THIRD OFFICERS' TRAINING SCHOOL, *Camp Upton, Long Island, New York, this* 26th *day of* March *one thousand nine hundred and eighteen.*

LIEUT. COLONEL, NATIONAL ARMY,
Commanding Officer
THIRD OFFICERS' TRAINING SCHOOL. Jsj

This certificate of graduation as a second lieutenant is the commission of Thomas A. Pyterman from the third officers training school at Camp Upton following the Plattsburgh model. Pyterman served in the 307th Infantry, Company E.

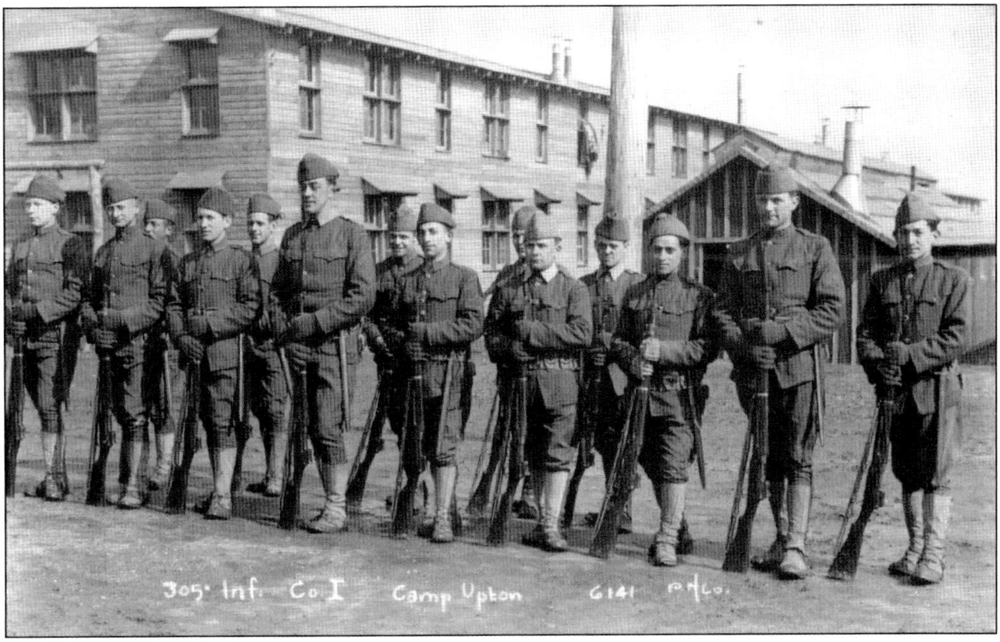

Wool overseas caps replaced campaign hats during the winter. This group from the 306th Infantry is drilling outdoors in the cold wearing wool hats and gloves.

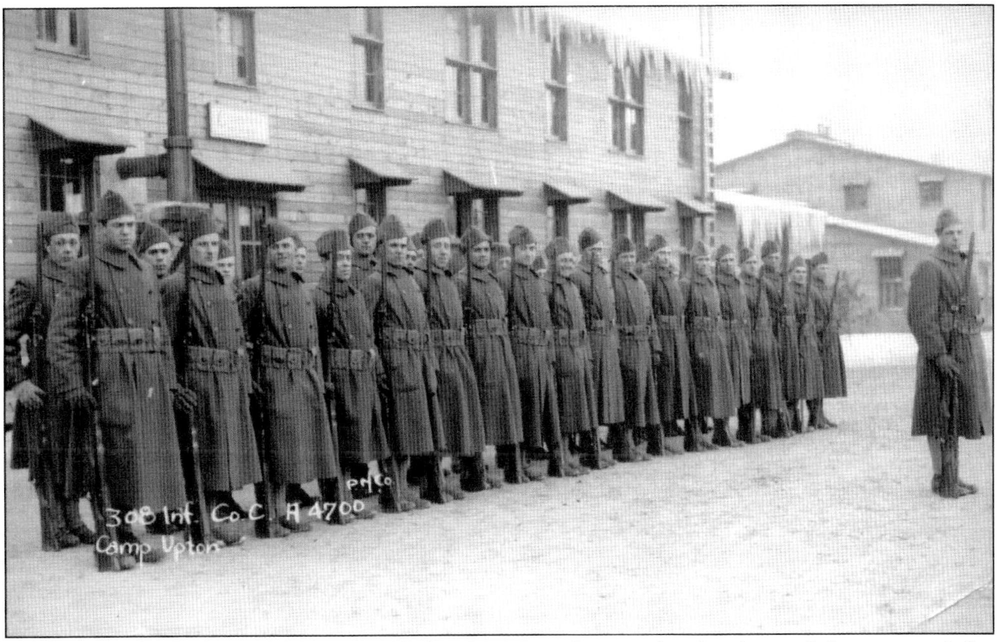

The winter of 1917–1918 was especially harsh, as can be seen by the dress of the men and the icicles on the barracks and latrine in this photograph. Training continued and intensified despite the cold and snow.

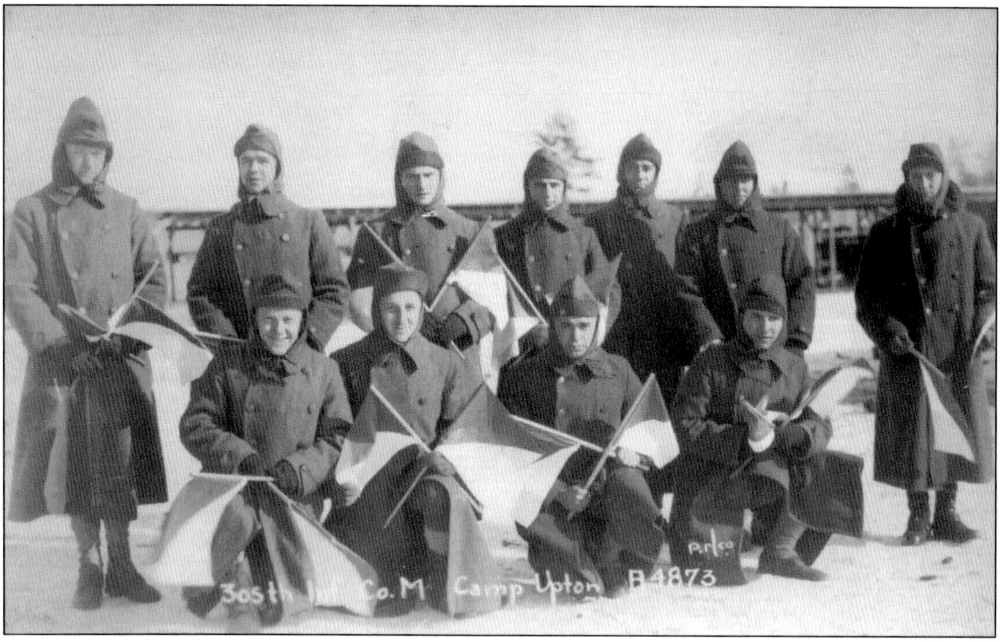

Orders and messages were transmitted by field telephones, signal flags, and runners. These soldiers of the 305th are taking a "camera break" from practicing two-arm semaphore in the cold and snow.

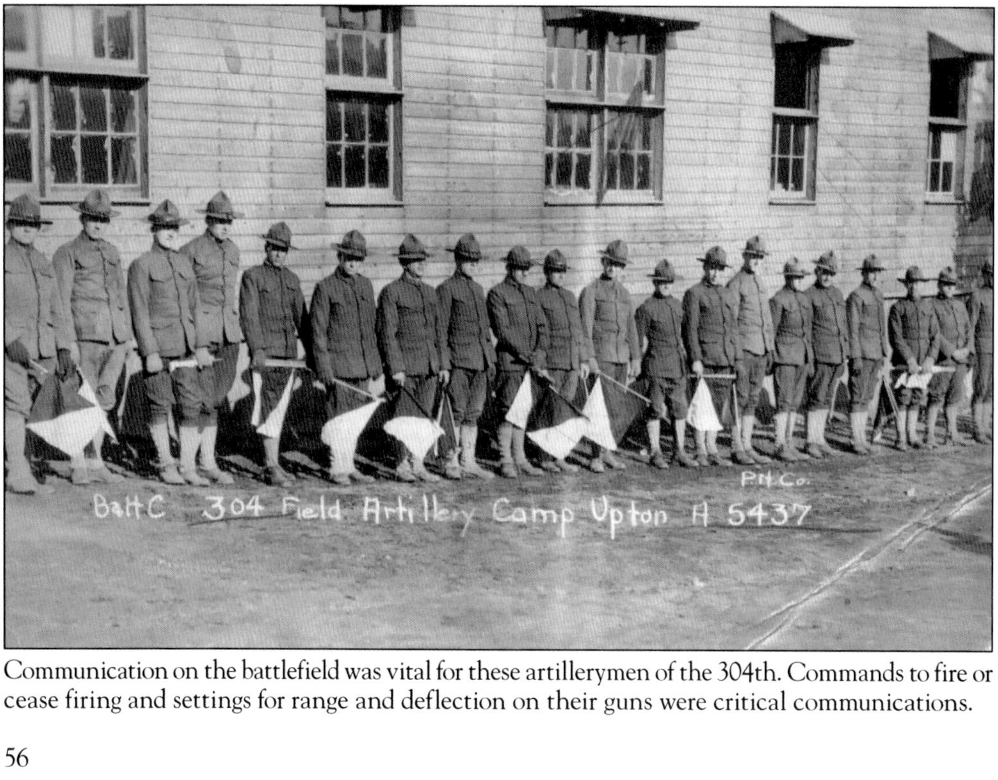

Communication on the battlefield was vital for these artillerymen of the 304th. Commands to fire or cease firing and settings for range and deflection on their guns were critical communications.

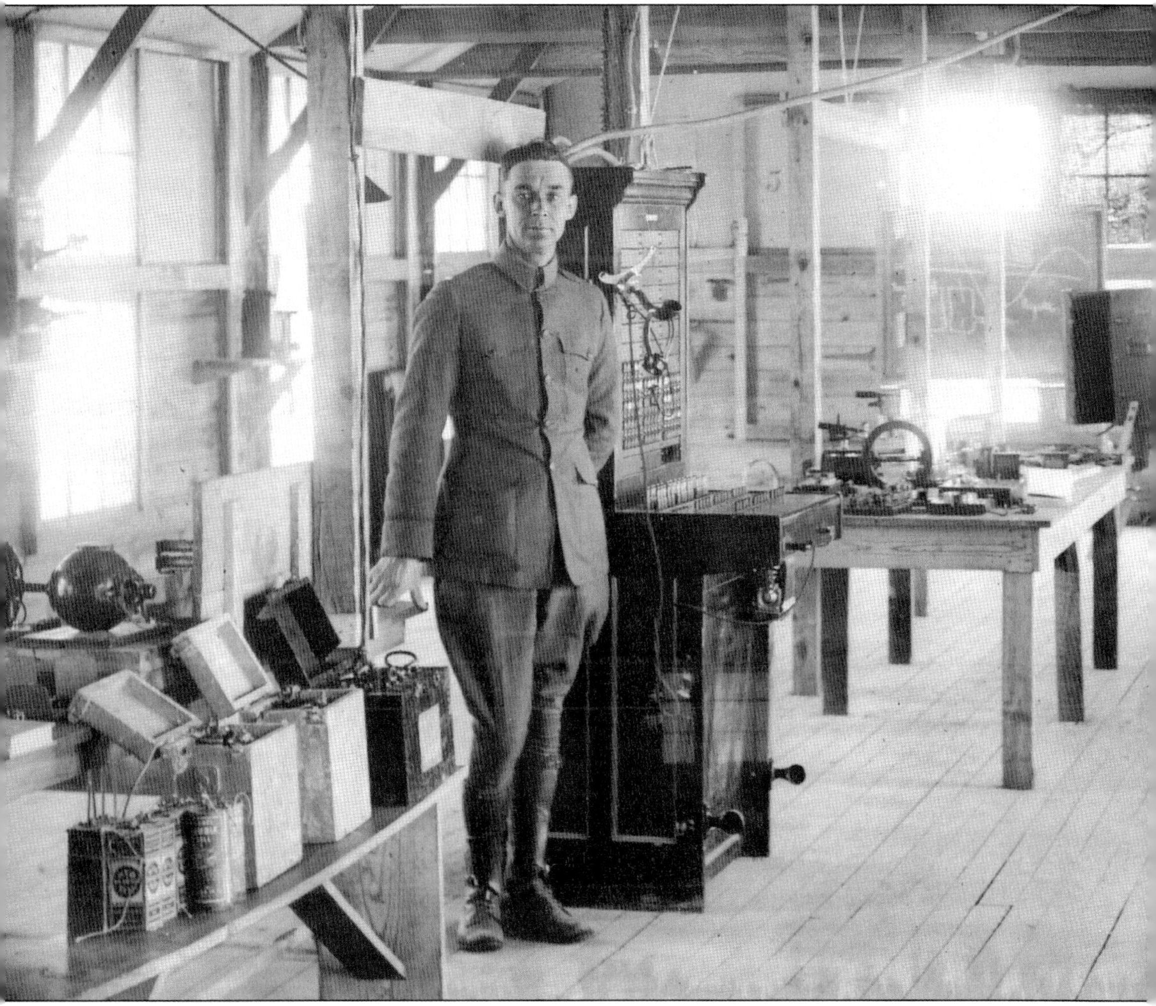

This officer from the 321st Field Signal Battalion stands in his old classroom. He wrote home about his experience that he was "working to whip our men into shape from an electrical point of view and make them able to get the messages through when we get 'over there.' "

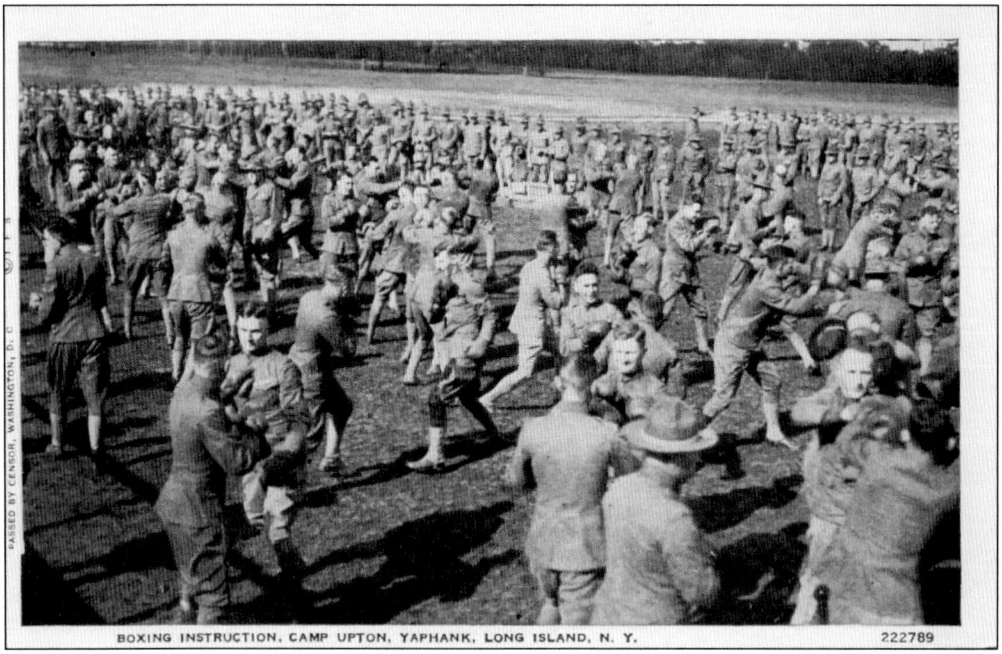

Boxing instruction was a basic part of physical exercise. It also conditioned the new soldiers to physical aggression and prepared them for hand-to-hand combat. Welterweight champion of the world Benny Leonard spent time at the camp training the men to box.

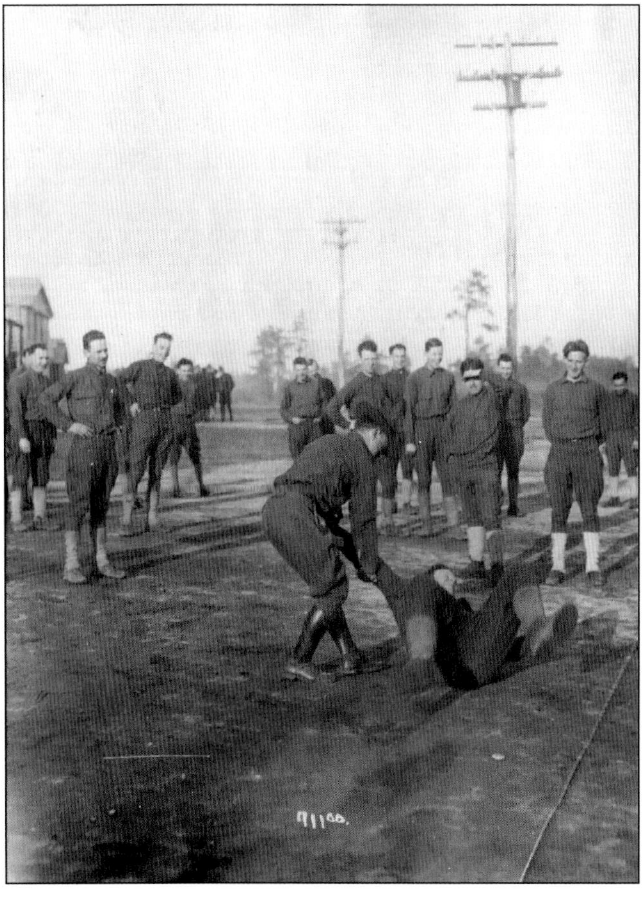

The men learned many forms of close combat. This officer (in the campaign hat) is teaching jujitsu.

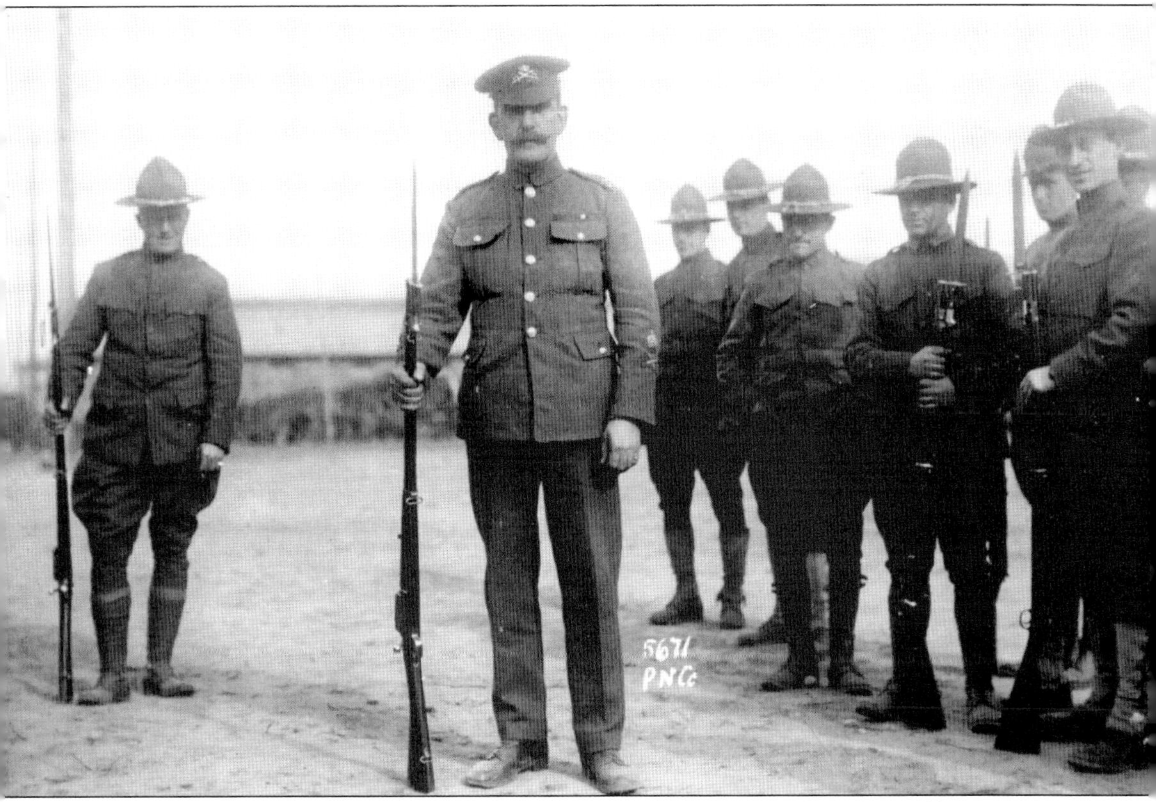

Sgt. Maj. G. Covington of the British Army demonstrates rifle drill. The rifles and bayonets are Krag-Jorgensens, which were replaced by the Springfield in 1903. Camp Upton received 7,000 of these older-model rifles on September 28, 1917.

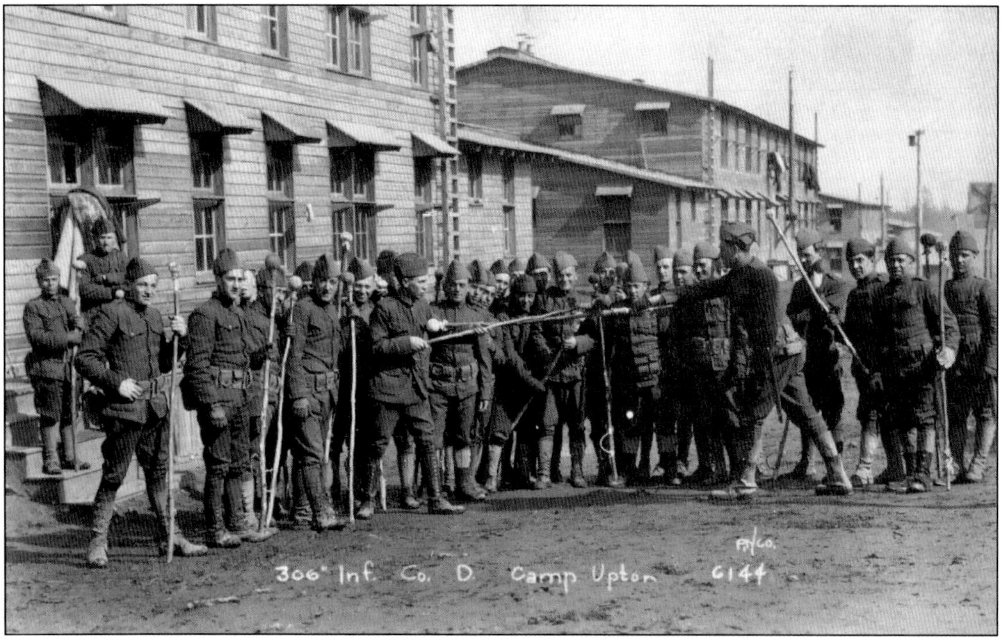

These men are practicing with poles instead of bayonets. Dummies were rigged up outside every infantry barracks for practice.

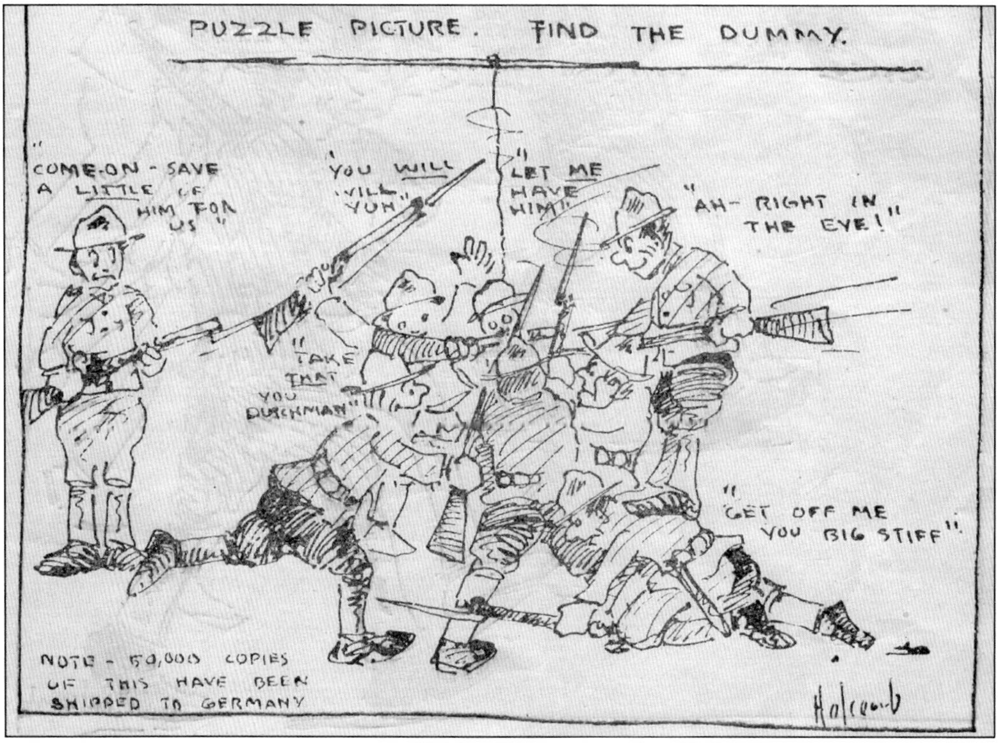

Drawing was encouraged in the camps. This cartoon appeared in a slim volume of drawings by Holcomb titled "Ups and Downs of Camp Upton," published in Brooklyn in 1917, which soldiers could buy for 15¢.

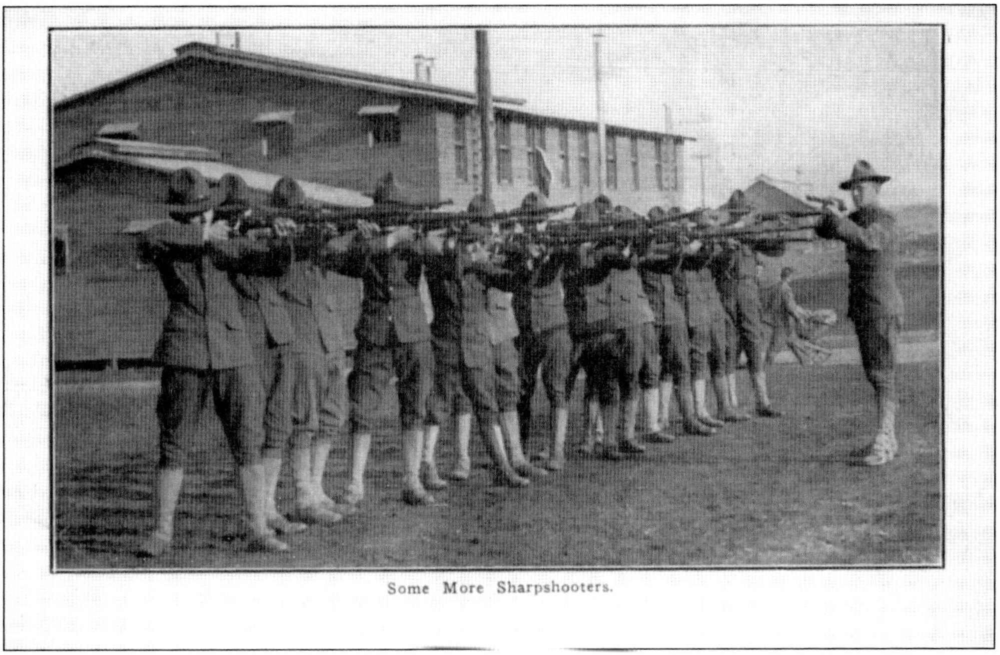

Krag-Jorgensen rifles were used before and during the brief Spanish-American War. They were given new life as drill rifles until enough new US Model 1917 rifles were manufactured to replace them.

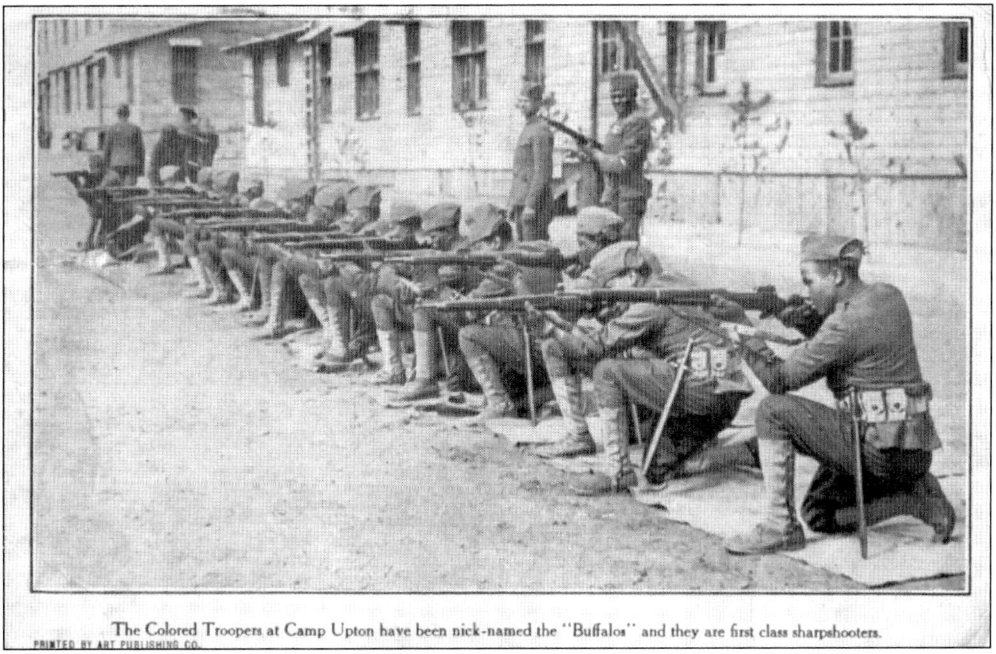

Black soldiers with US Model 1917 rifles practice firing from a kneeling position. They are most likely members of the 367th Infantry, 92nd "Buffalo" Division, which was organized at Camp Upton on November 3, 1917. Several thousand black troops trained at Camp Upton in segregated barracks.

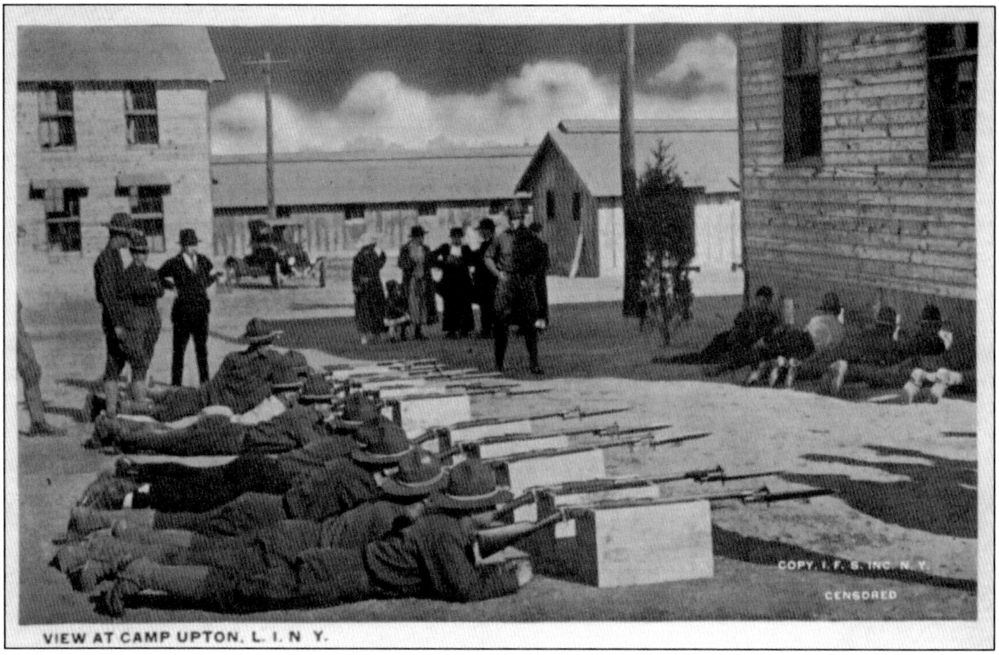

American generals insisted that US troops be trained in marksmanship and open field maneuvering. Recruits learned to use a rifle in individual rifle practice. Here they are learning to use and adjust the sights, as visitors to the camp look on.

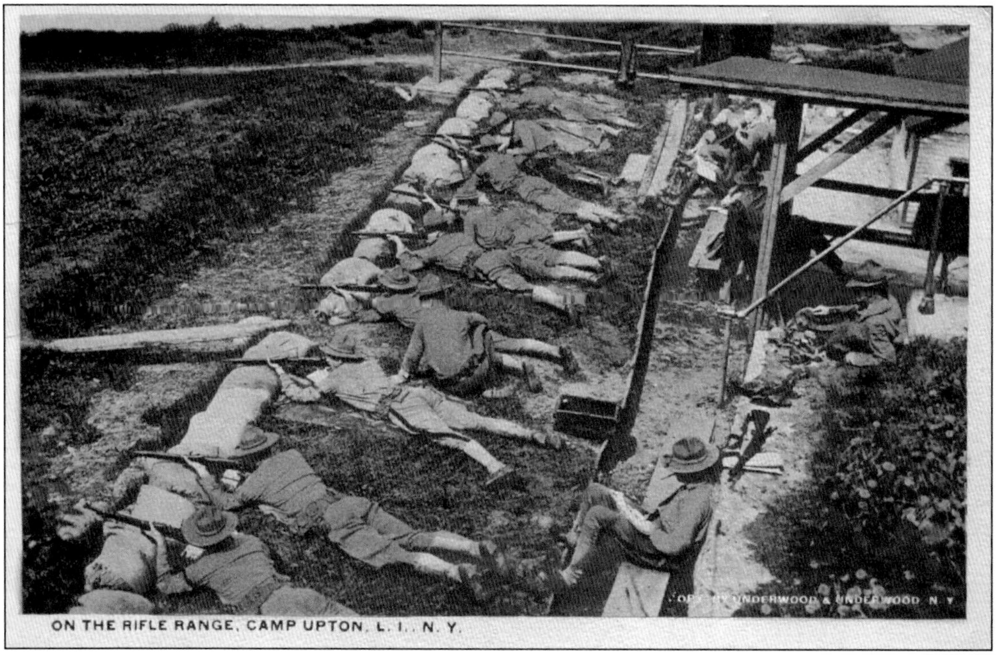

The rifle range was completed and opened in December. Live fire instruction began immediately. Every soldier was required to complete a training course involving firing 150 rounds.

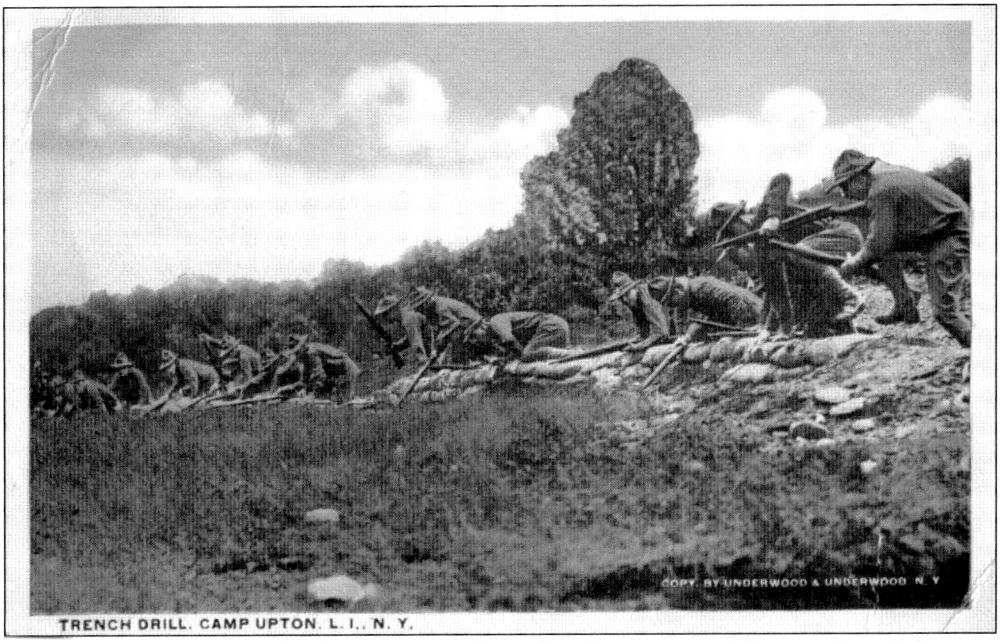

These soldiers are getting familiar with trench warfare, leaping out of the trench and charging with bayonets fixed to attack the enemy in the opposing trench. The distance could range from a few hundred feet to several thousand feet, and crossing it resulted in many casualties.

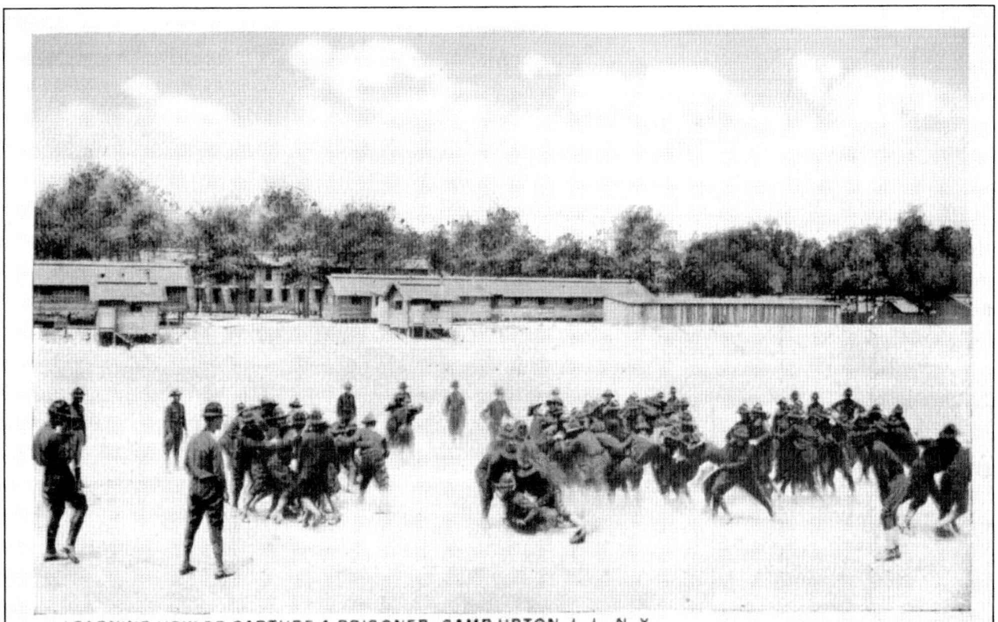

Enemy soldiers were a valuable source of information, and typical trench warfare included raids on enemy outposts and trenches to obtain prisoners. These soldiers are learning to overpower and capture an enemy.

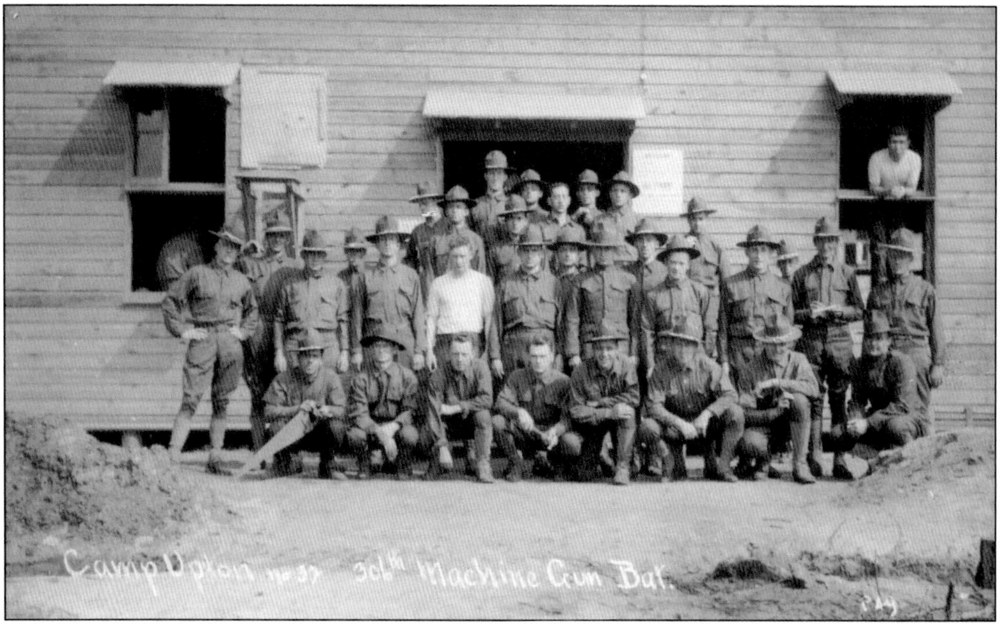

The soldiers of the 306th Machine Gun Battalion relax outside their barracks. There were three machine gun battalions in the division. They were well trained, but American machine guns were obsolete. The American Expeditionary Forces took no machine guns when they went over. Once in France, they were given French Hotchkiss machine guns and trained to use them.

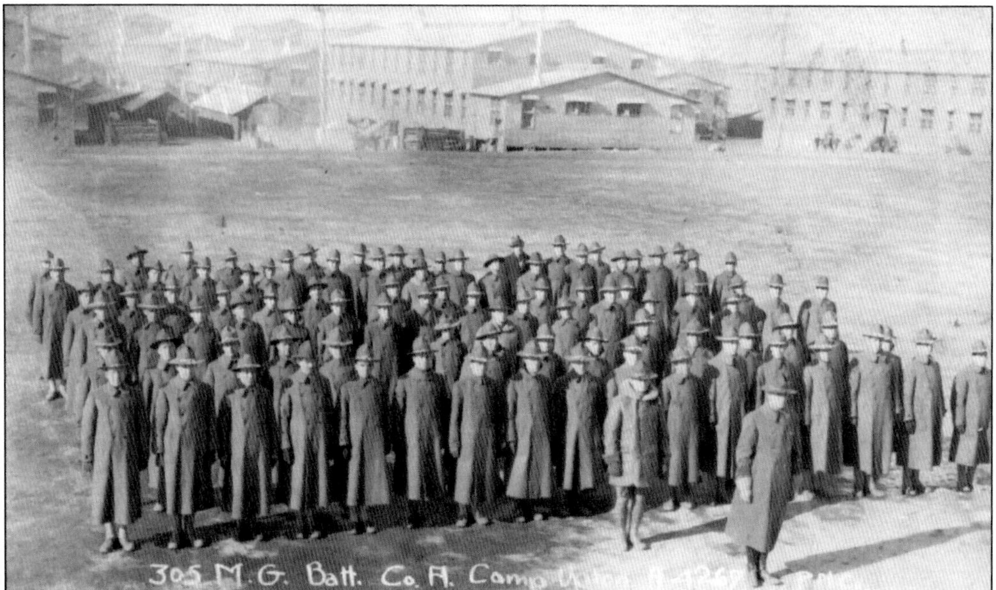

Company A, 305th Machine Gun Battalion, lines up on the parade ground. They are wearing their new winter overcoats but have retained their campaign hats. A full-strength machine gun company included about 175 officers and men. Each company had 16 machine guns: 12 to use and 4 spares.

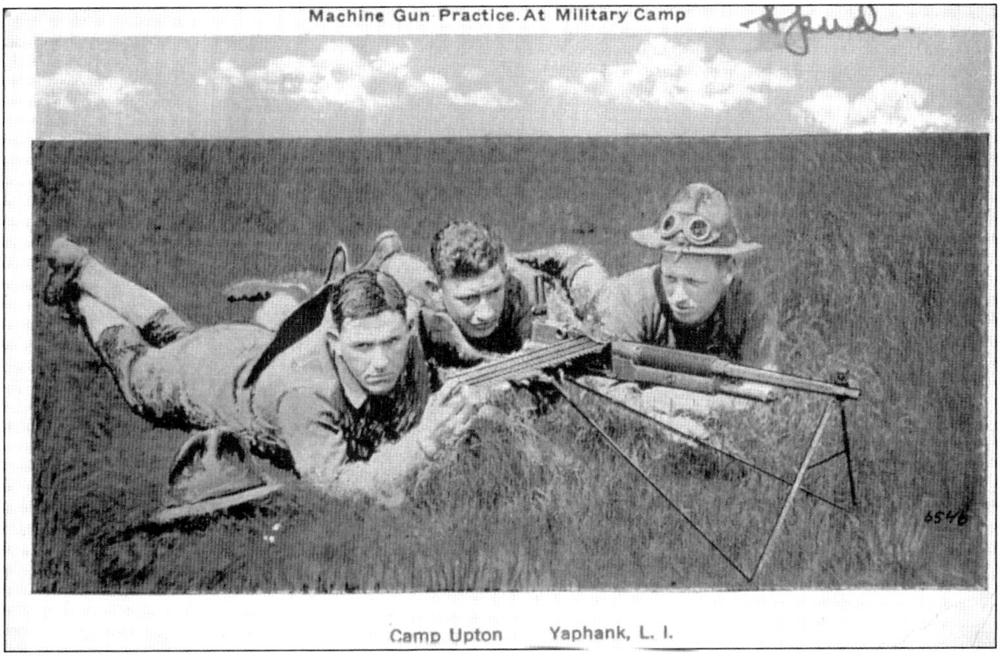

The Hotchkiss M 1914 machine gun with its crew of three is shown here. A squad of nine gunners was assigned to each gun to serve as replacements for the crew firing the gun and to keep a large supply of ammunition close at hand. Their equipment included the ammunition cart needed to transport the required supply of 30-round magazine strips.

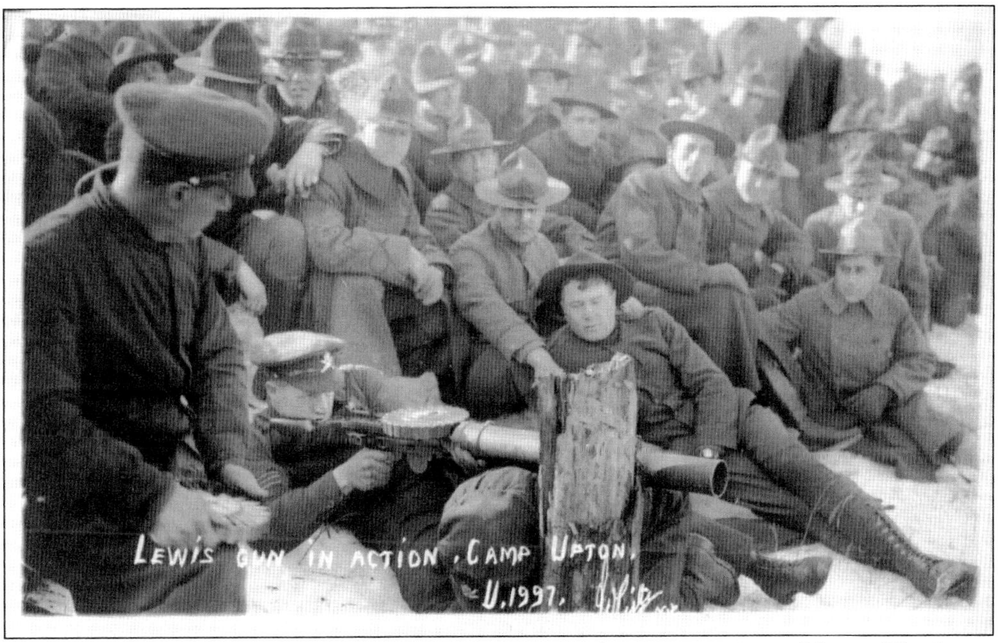

A British officer and sergeant demonstrate a Lewis machine gun at Camp Upton. The Lewis was the standard light machine gun in the British Army, but it was not used by the gunners of the 77th Division.

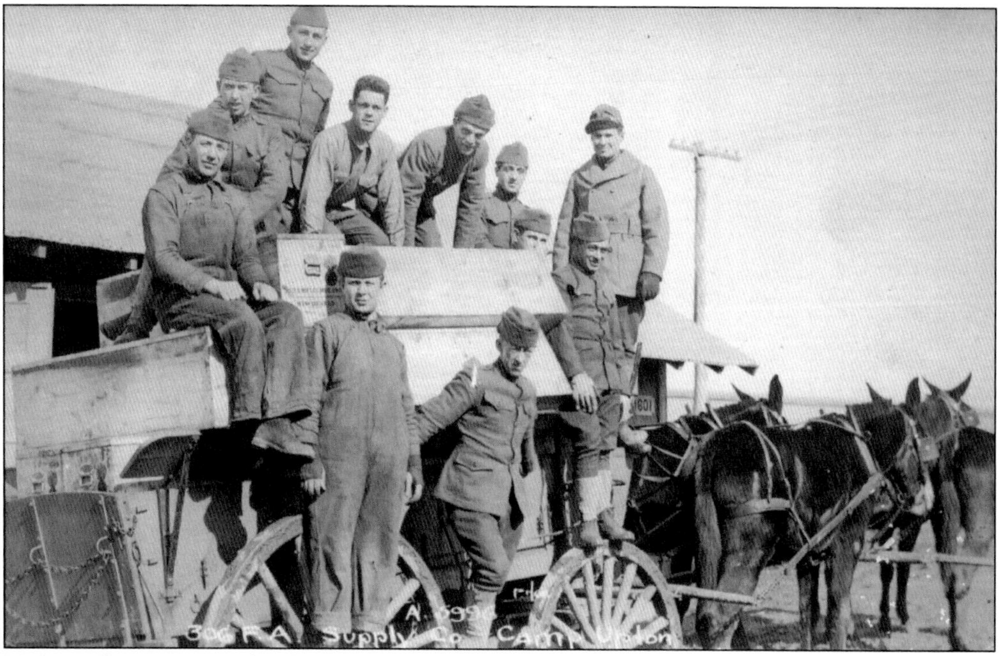

Mules pulling a wagon for the Supply Company of the 306th Field Artillery finally deliver new rifles. It proved impossible to produce enough of the Springfield Model 1903 rifles adopted by the US Army for all of the new divisions. It was decided that the American factories producing Enfield rifles for the British Army in .303 caliber would modify their equipment to produce the rifle to use the American .306 ammunition. These rifles were called the US Model 1917. The rifles in the crates are the best of the new US Model 1917 rifles, made by Winchester. They replaced the obsolete Krags.

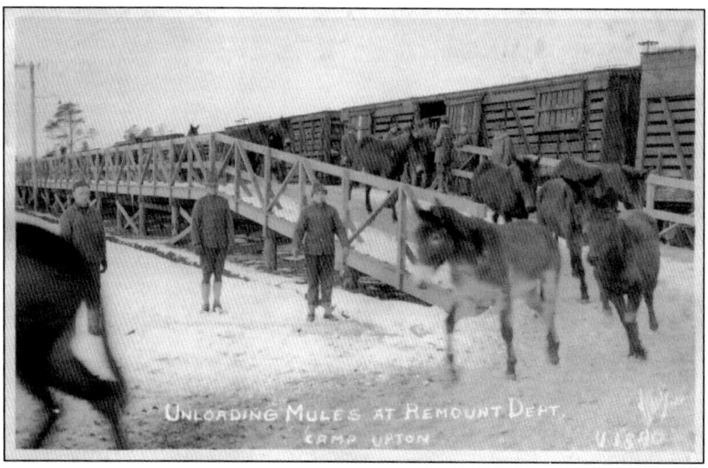

The mules finally arrived in December 1917. They would take on the task of hauling supplies and the artillery around the camp. Each field artillery battery required six horses to pull each limber attached to a gun or a caisson, which was used to transport ammunition and other supplies. A battery could require about 150 horses.

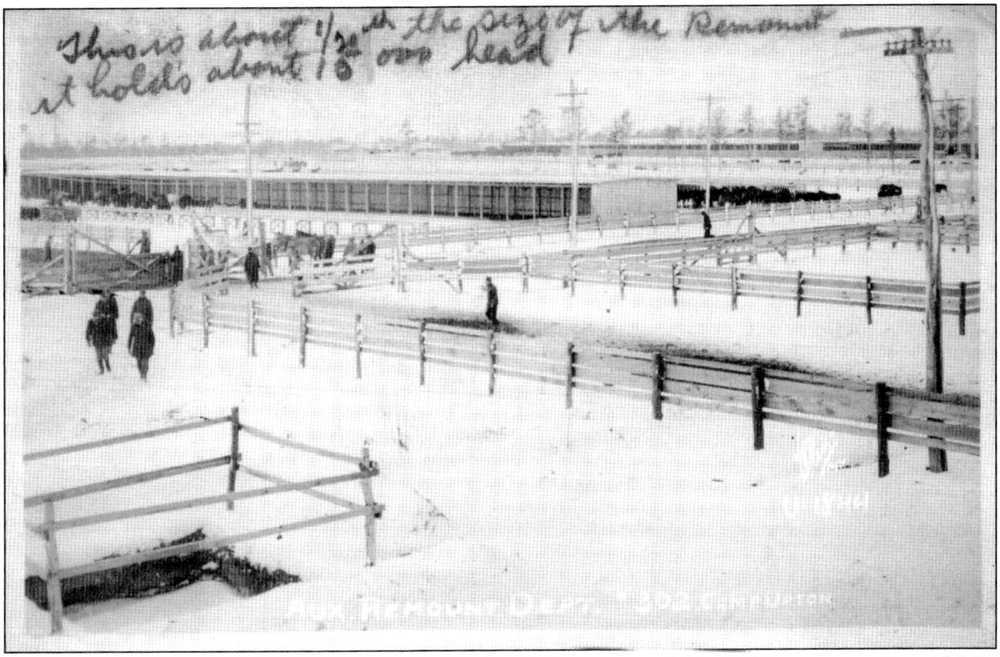

Eighty-seven horses arrived on December 10. The remount or stables could accommodate 15,000 horses in 312 buildings.

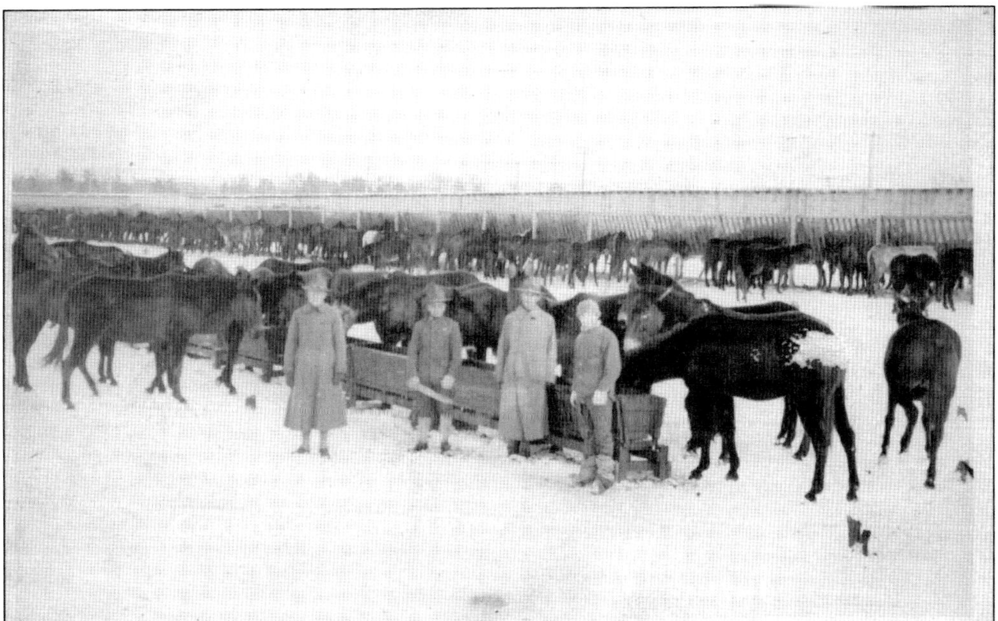

Horses are fed at Camp Upton. The horses consumed 27,720 tons of hay and over a million bushels of oats.

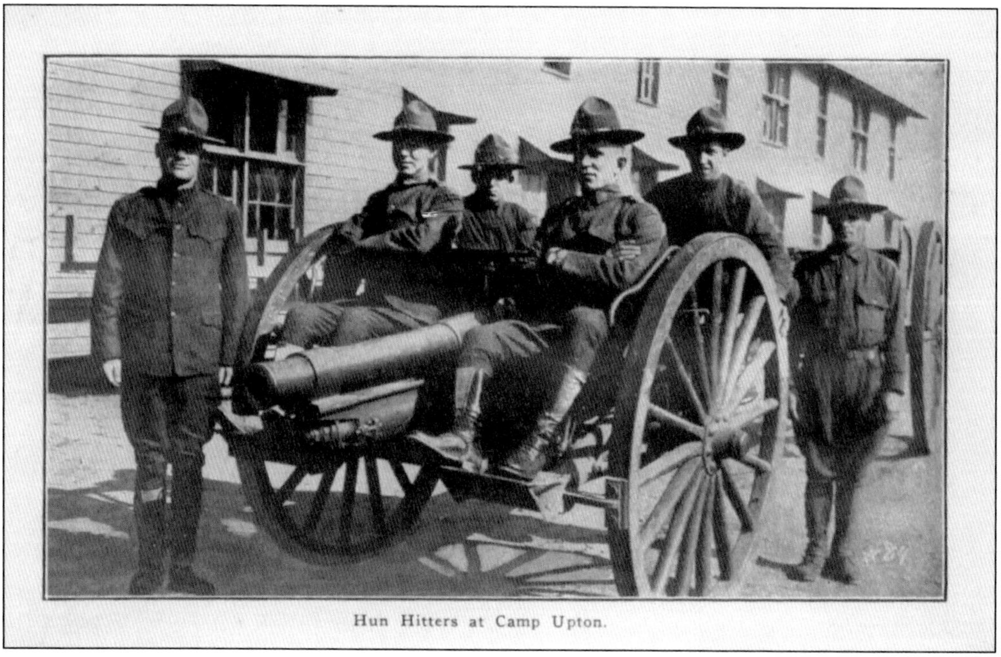

In this photograph, Camp Upton artillerymen of the 304th work with a Model 1902 three-inch field gun. These guns were used for training only and were not taken to France. US field artillery regiments were equipped with the legendary French 75-millimeter guns during their training period in France and throughout the war.

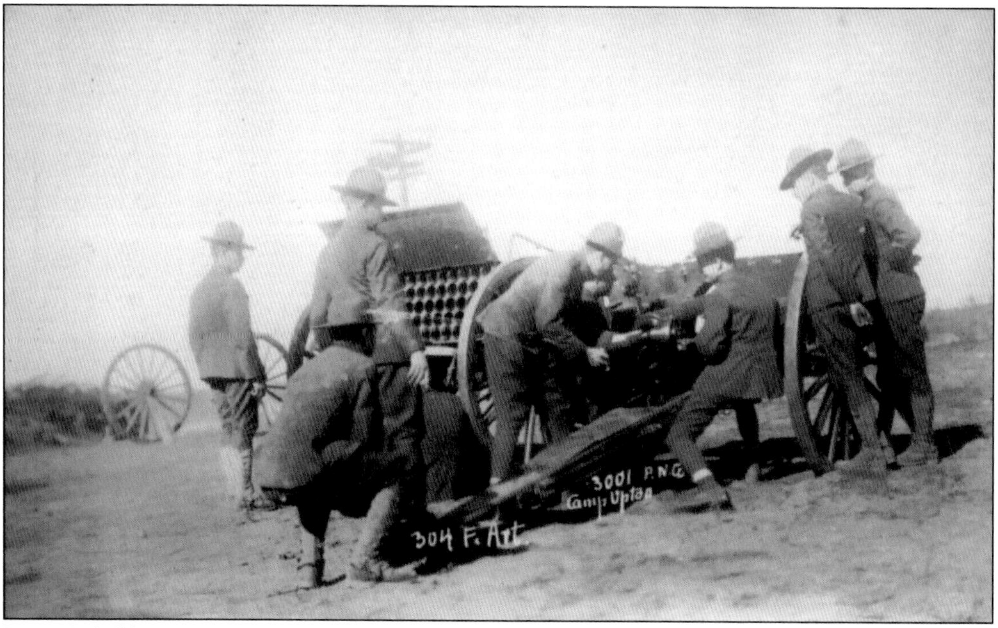

In 1917, field guns were in very short supply in the United States. Camp Upton had none until Henry L. Stimson, former US Secretary of War, who was a newly assigned officer in the 305th Field Artillery, obtained a battery of four guns from the New England National Guard on November 10. These guns were shared with the other regiments in the division for dry-fire drills.

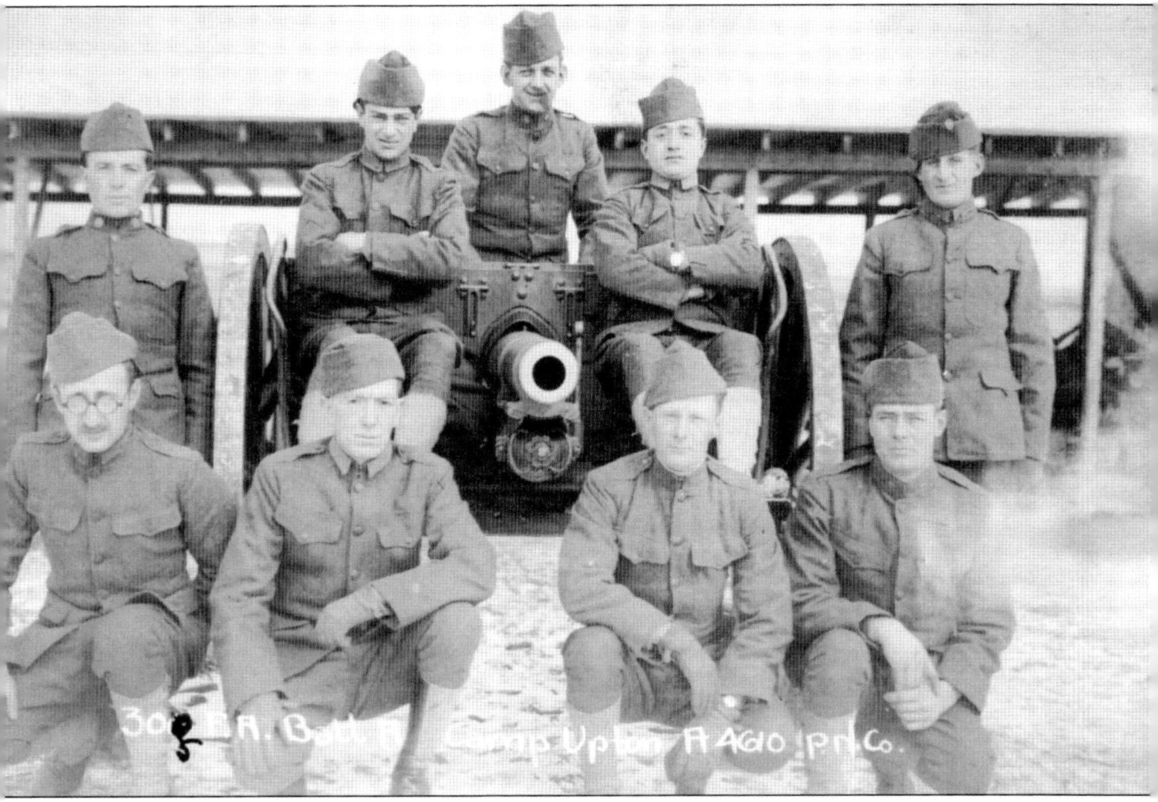

On December 12, 1917, the 305th fired 19 live rounds on the Camp Upton range. It fired 10 more on December 13. This was the only live firing by the field artillery of the 77th Division before training began in France.

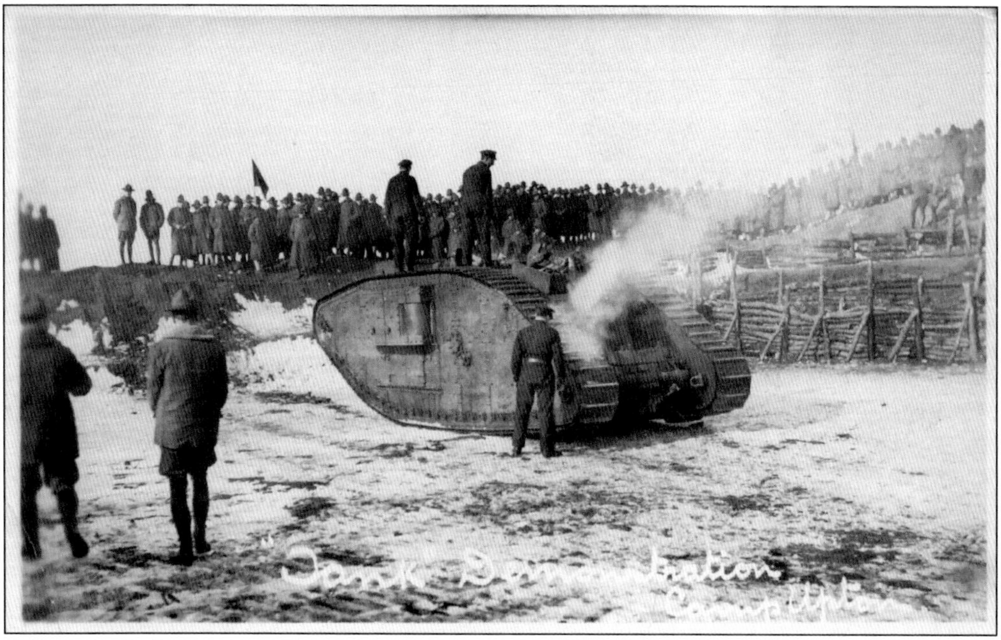

A British Mark I tank was brought to Camp Upton in December 1917 and demonstrated for the entire division. This was the first modern tank; they were first used as a surprise weapon on September 15, 1916, in the Somme Offensive.

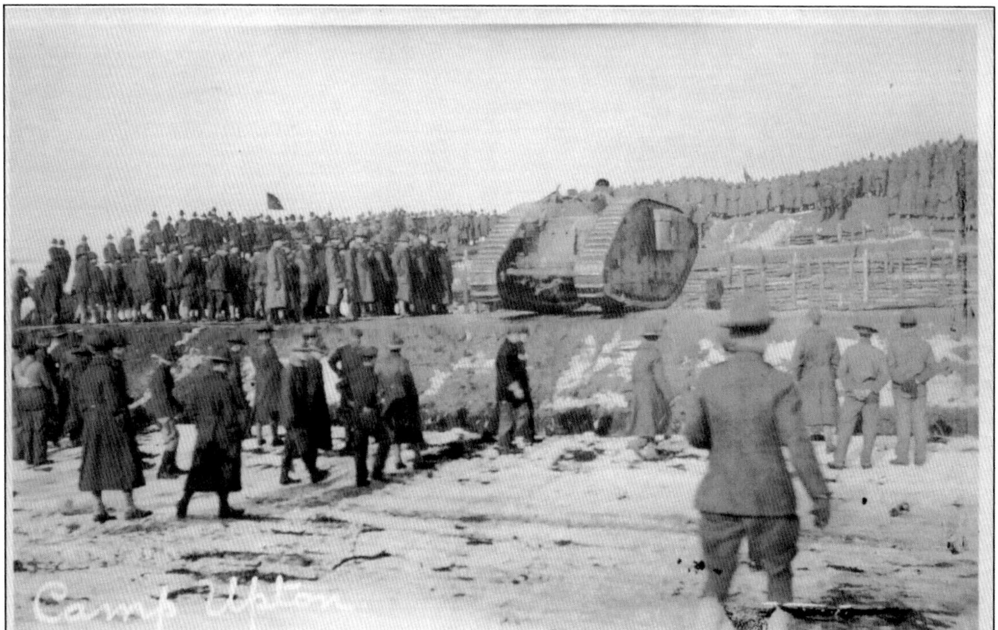

The British developed tanks in a secret program to break the stalemate of trench warfare. The first British model could crush barbed wire barriers, cross trenches, and withstand rifle and machine gun fire.

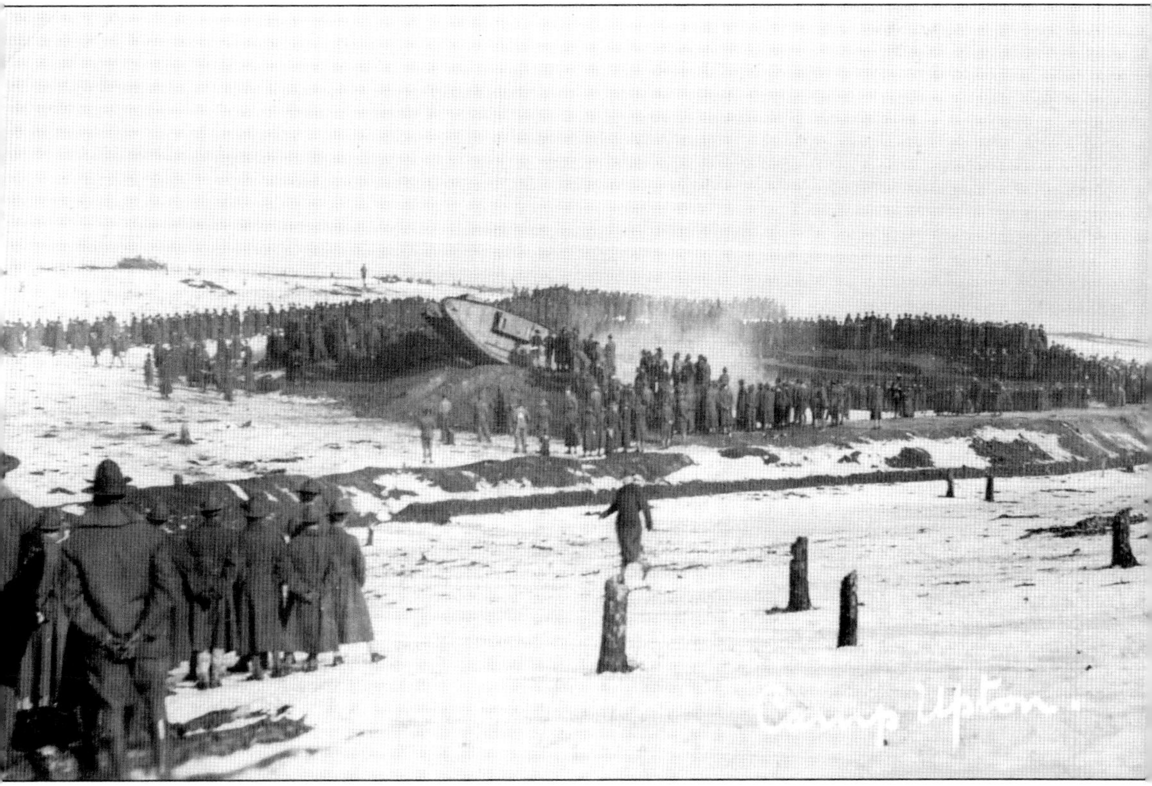

This tank, named *Britannia*, was demonstrated to the officers and enlisted men of Camp Upton in the practice trench area. The tank was propelled by a six-cylinder gasoline engine and carried several machine guns.

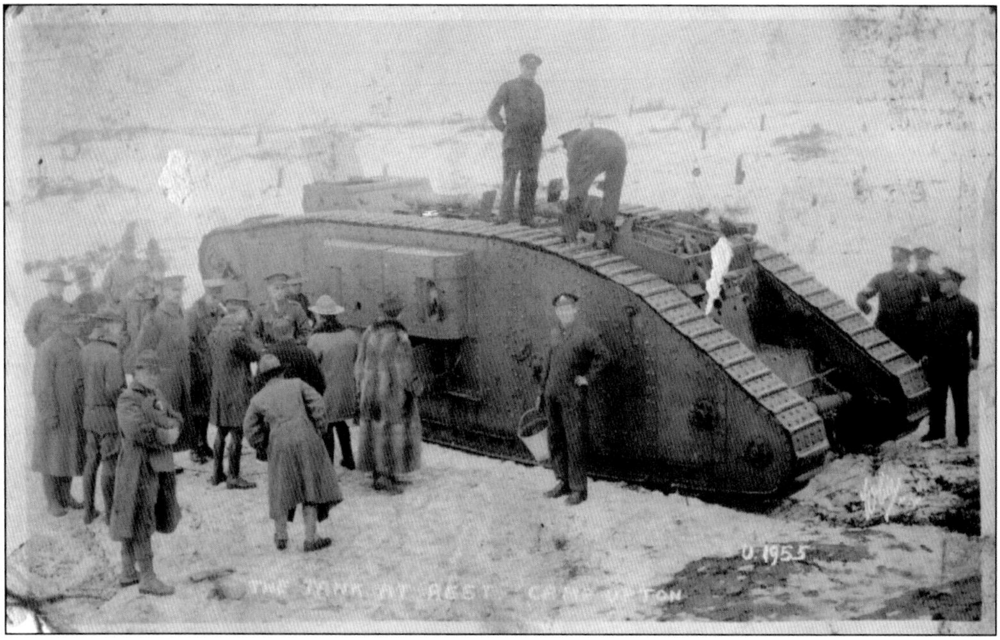

The tank was propelled by caterpillar treads similar to those used on tractors, which had been developed and patented by an American company, the Holt Corporation, in 1910. This model could carry a six-pounder cannon and several machine guns.

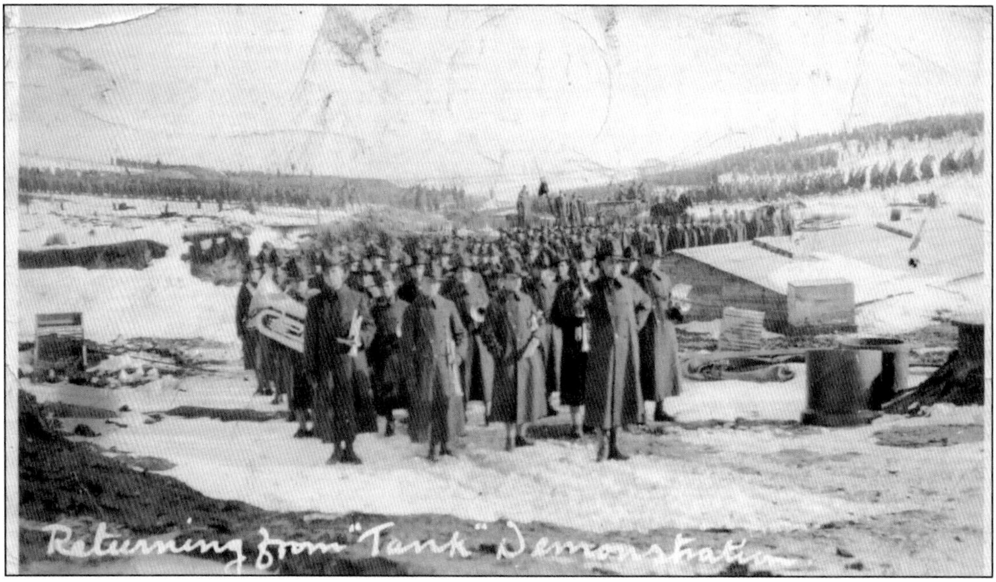

Training at Camp Upton included the use of music to reinforce discipline and encourage the troops. The march returning from the tank demonstration pictured here is an example of this.

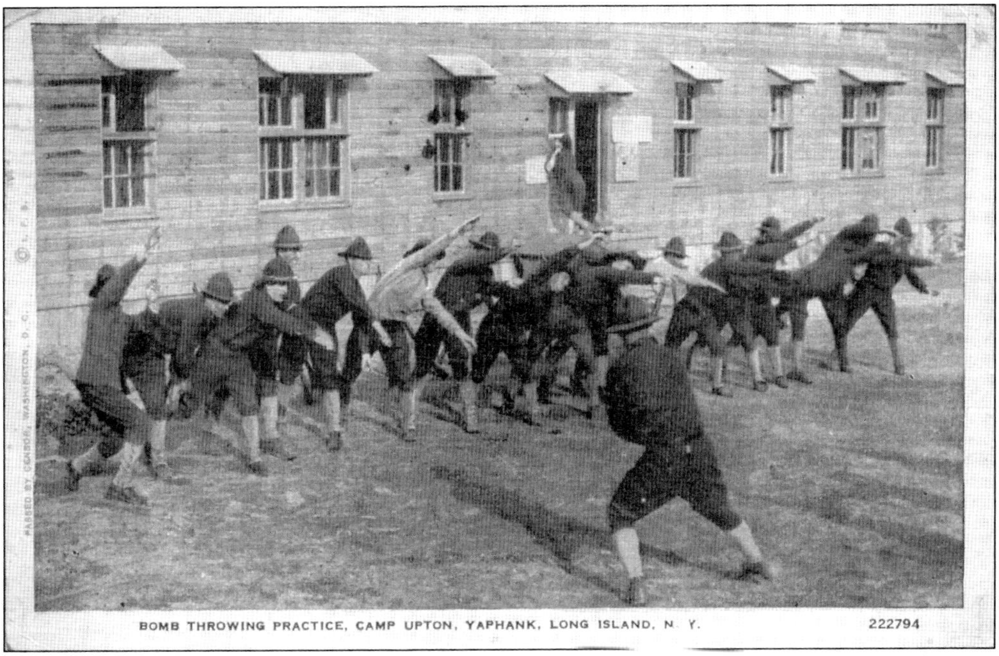

Hand-thrown bombs, better known today as hand grenades, were very effective in clearing trenches of enemy soldiers. These soldiers are practicing the proper technique for throwing in front of their barracks.

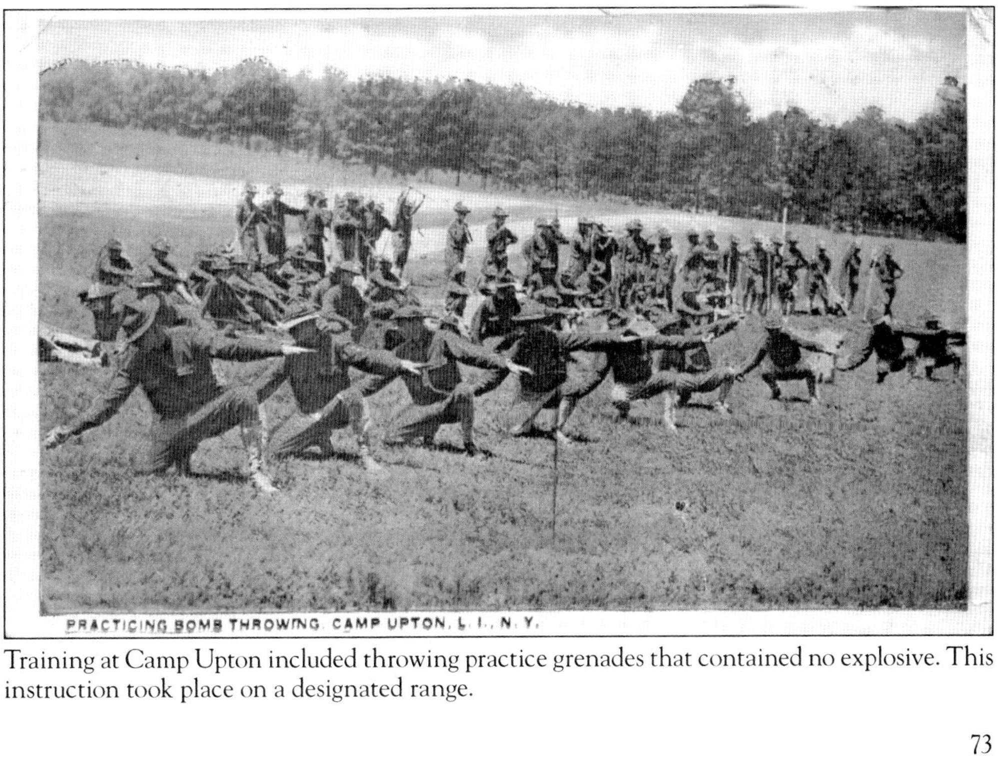

Training at Camp Upton included throwing practice grenades that contained no explosive. This instruction took place on a designated range.

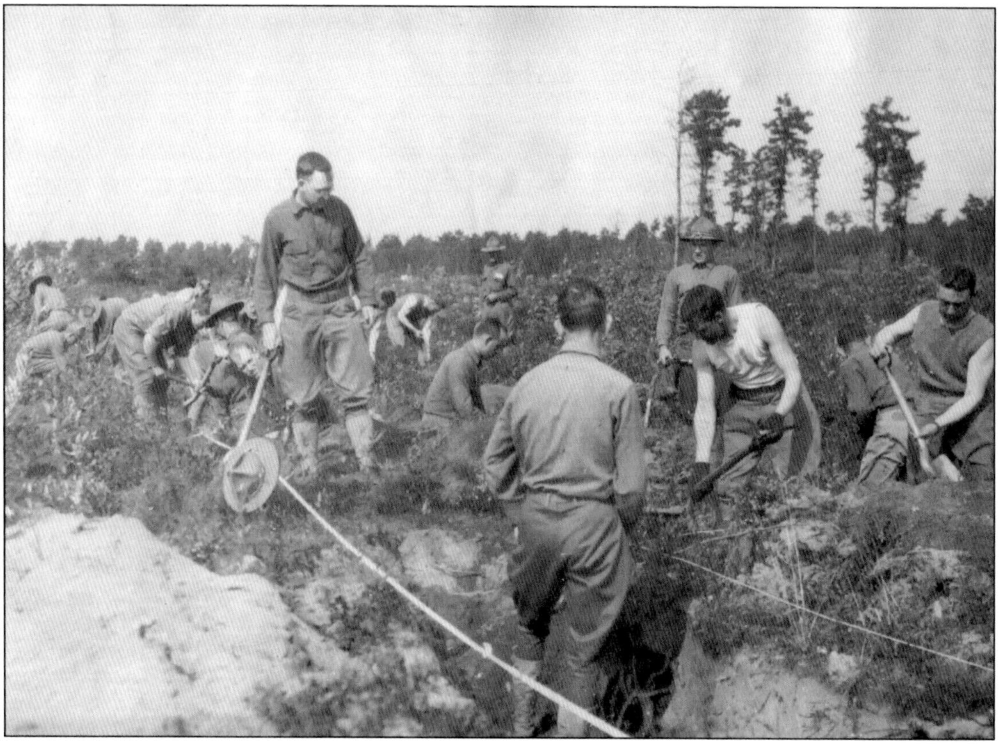

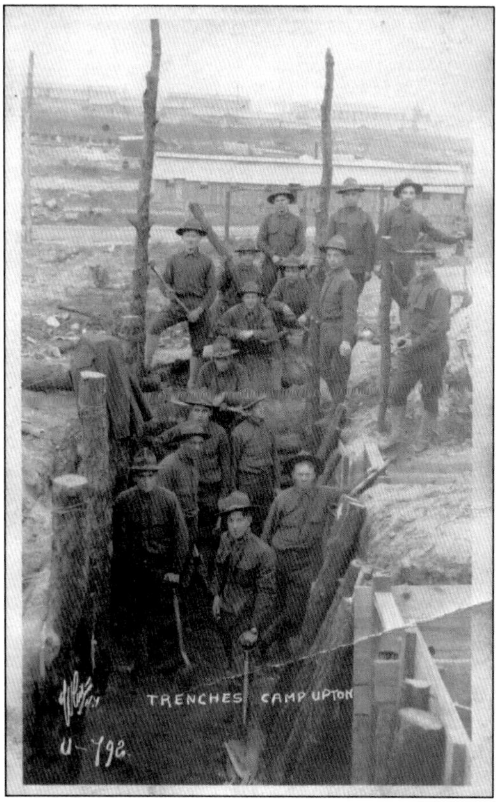

In November 1917, members of the 302nd Engineers laid out and supervised the building of the trench system so that soldiers could experience the conditions they would encounter in France. When the site was surveyed for a New York State Parks Historic Sites inventory in 1981, several thousand feet of the original trenches remained. More have been discovered during various construction activities and have been preserved as historic landmarks.

The earthwork training trenches were designed to duplicate European trench systems. They were dug by soldiers by hand. Trench warfare infantry assault tactics emphasized preparatory use of artillery and gas.

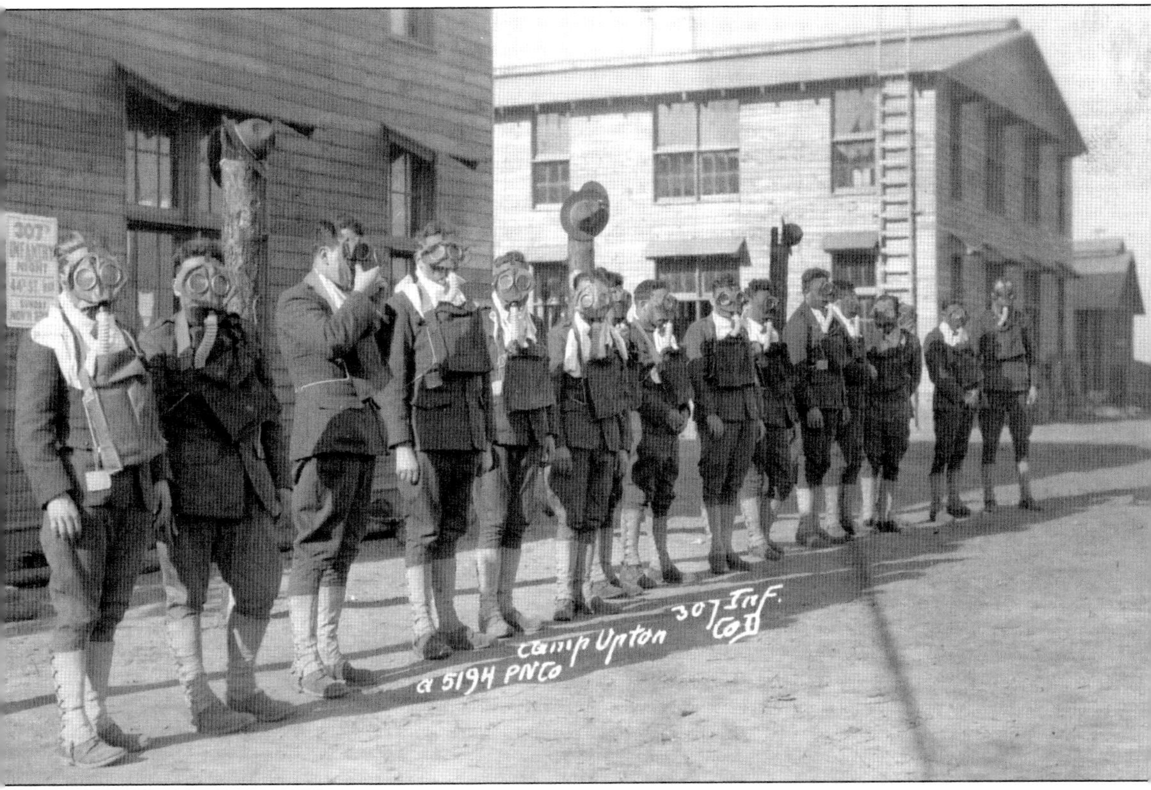

Soldiers were trained in new methods of modern combat, including gas and trench warfare. "Mustard gas" was first deployed by German forces against British troops in 1915. Exposure was feared, and early gas masks only protected the lungs, not the skin, which developed painful blisters and caused blindness and death. On March 21, 1918, Cpl. Oscar Ostrow wrote in his diary, "I attended gas school and we had a lecture first. Then we went to the gas house and were taken through the chlorine and tear gas which convinced me that the masks were absolutely gas proof. We were also given a sniff of chloropicrin gas, mustard gas, and phosgene gas. We were then taken and shown through a dugout."

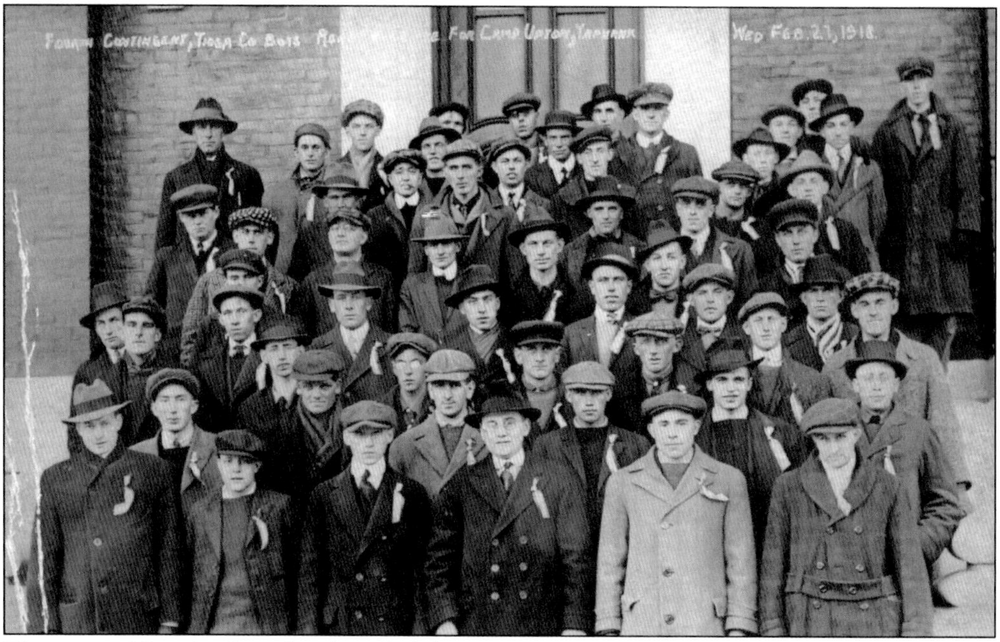

The story of Stanley Popielowski, a Polish immigrant on Long Island, could be that of any of the soldiers pictured here in the "Foreign Contingent." Popielowski was 26 years old when he enlisted on June 24, 1918, in his adopted hometown of Glen Cove. One of many immigrants who volunteered, he participated in the Meuse-Argonne Offensive of October 6–11, 1918. He received $89.90 when honorably discharged as a private at the Evacuation Hospital at Camp Upton on July 1, 1919, including a $60 bonus. He was not naturalized until July 26, 1920.

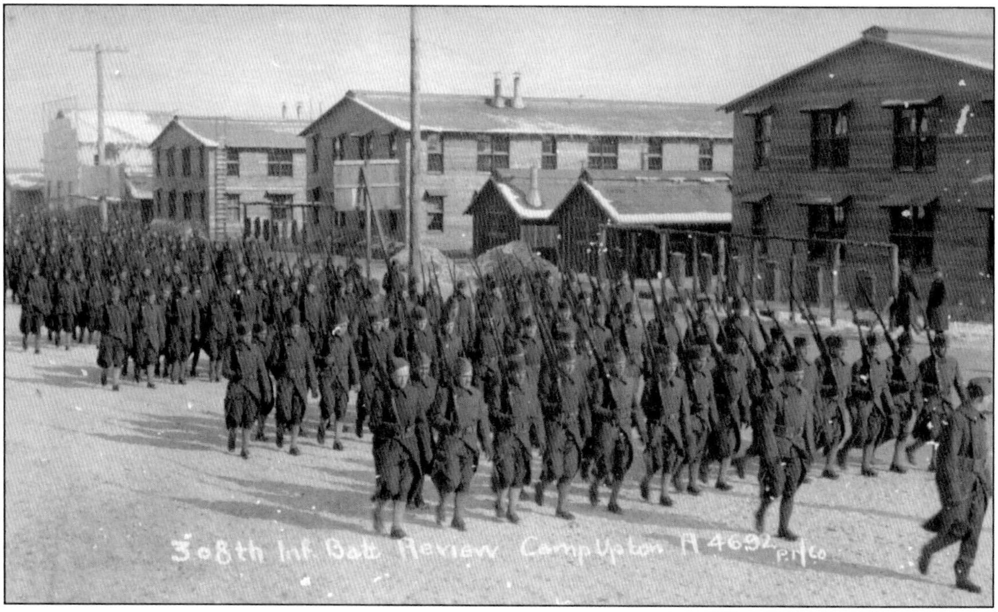

Shown here is a battalion of the 308th Regiment passing in review. A battalion consisted of four infantry companies and was commanded by a major. Three infantry battalions and one machine gun battalion comprised a regiment, commanded by a colonel. A 308th Regiment battalion review is shown here.

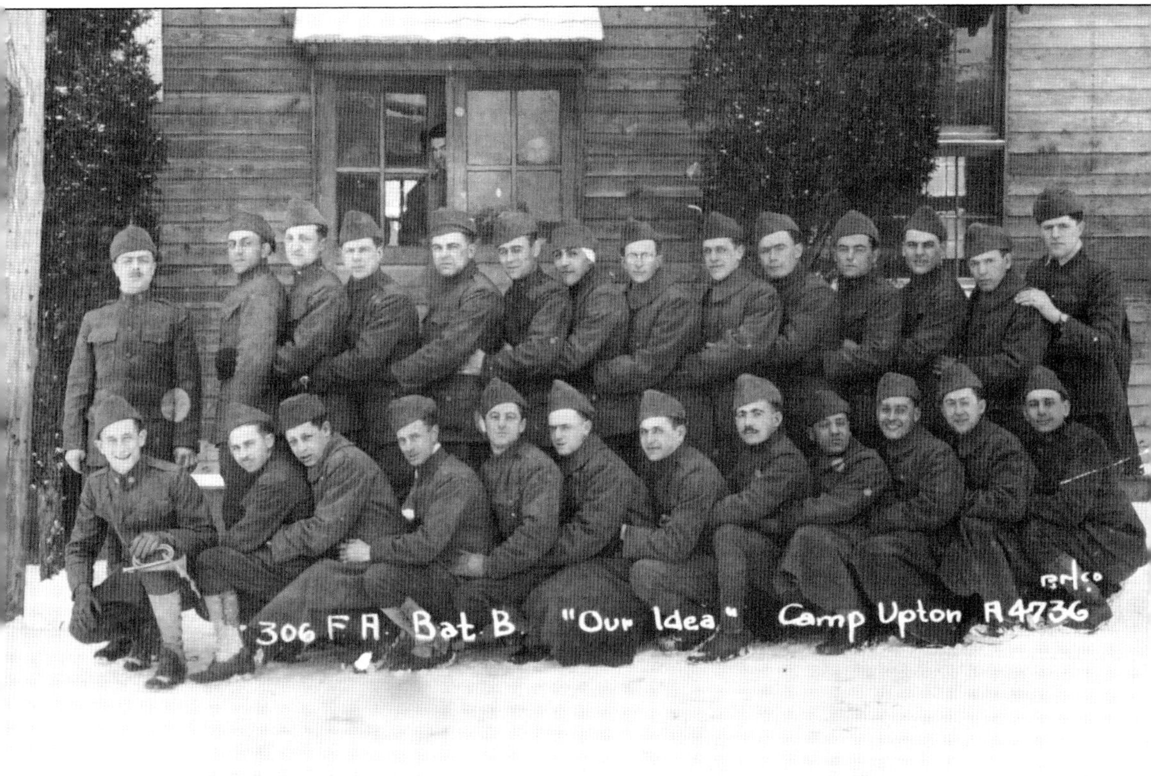

A rumor swirled in January 1918 that the Army would be breaking up the 77th Division and would use its soldiers as replacements in France. Most likely, these comrades chose to express "our idea" in this photograph that they had trained together, lived together, and they were staying together.

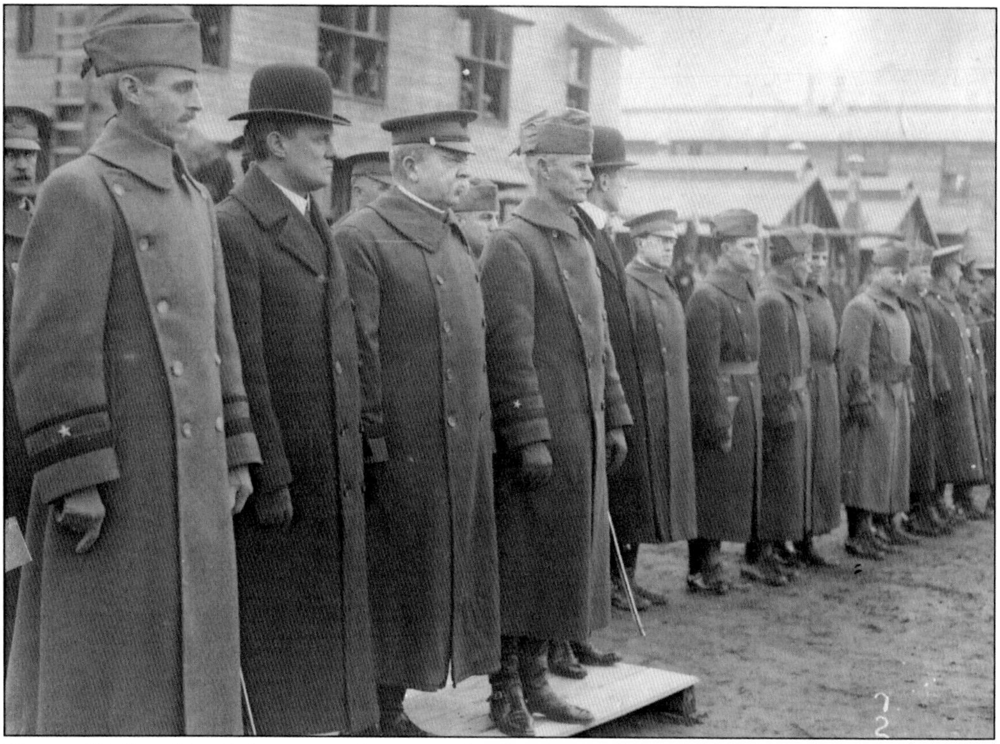

General Bell and his fellow generals review the division as departure for Europe gets closer. Bell traveled to France in December 1917 to survey battlefield conditions and returned in March 1918. However, he failed his physical examination and did not accompany his division to France.

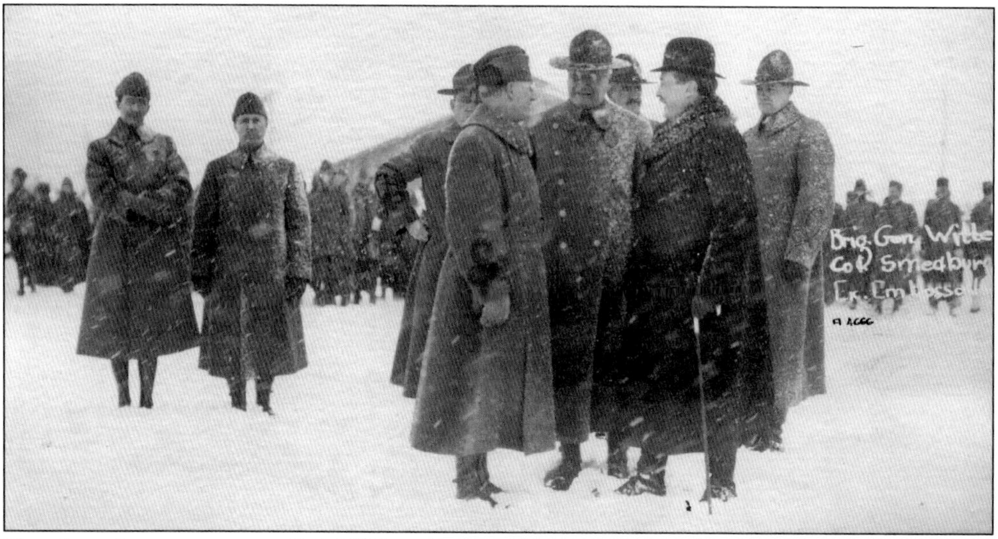

Noted machine gun expert Col. William R. Smedberg Jr. is shown here (with mustache, behind General Bell at center) with Brig. Gen. Edmund Wittenmyer and ex-ambassador to Germany James W. Gerard reviewing operations at Camp Upton in the winter of 1918. In 1917, Colonel Smedberg served as the senior instructor at the officer training camp at Madison Barracks in Sacketts Harbor, New York. In September of that year, he activated the 305th Infantry Regiment at Camp Upton and trained its personnel of 3,600 men.

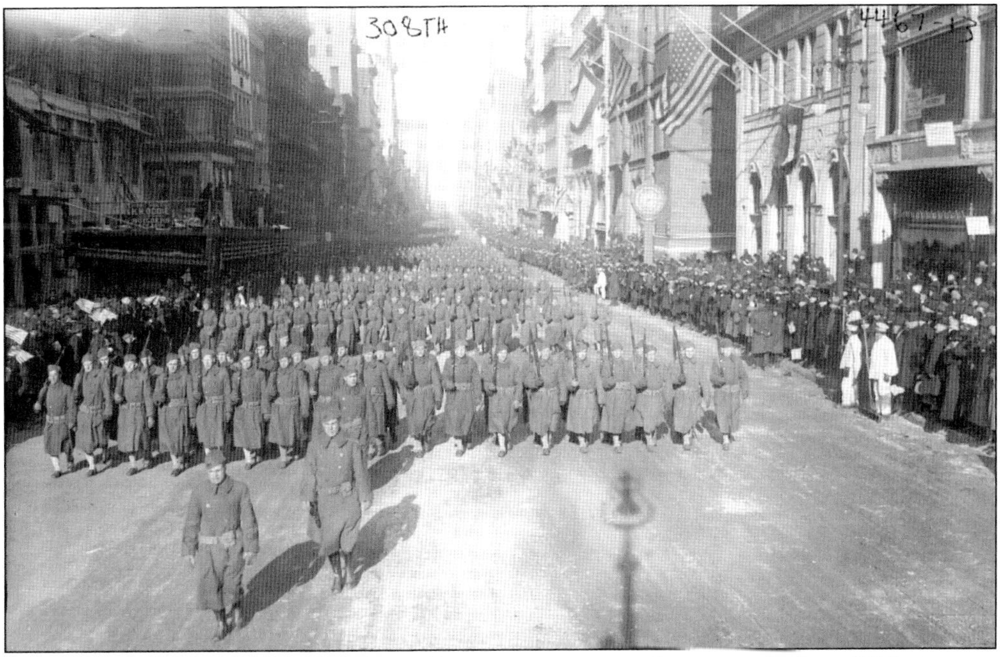

On February 4, 1918, the 308th Infantry Regiment participated in a parade down Fifth Avenue. This photograph was published in the *New York Times* with this caption: "Camp Upton men on review here: Eager soldiers of the 308th Infantry show results of two months training. All look physically fit. Bitter cold thins crowd of spectators. Governor and Mayor in reviewing stand."

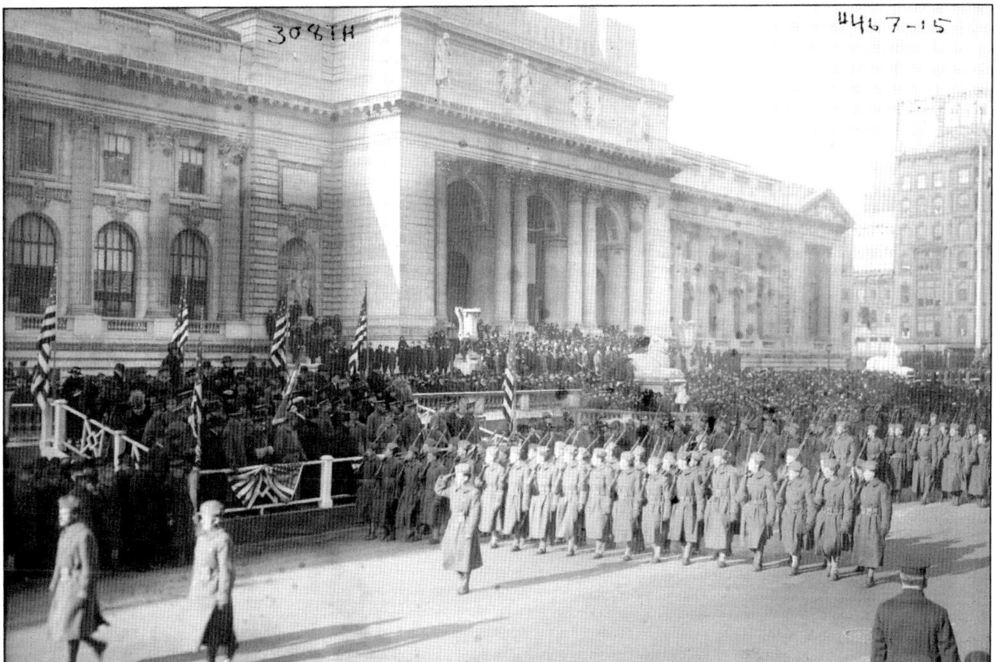

Maj. J.O. Adler reported in the *History of the 77th Division* that during the parade, a presentation of colors was made to the 308th at the New York Hippodrome. This regiment was "the first National Army unit to receive its colors and to parade before the people of New York."

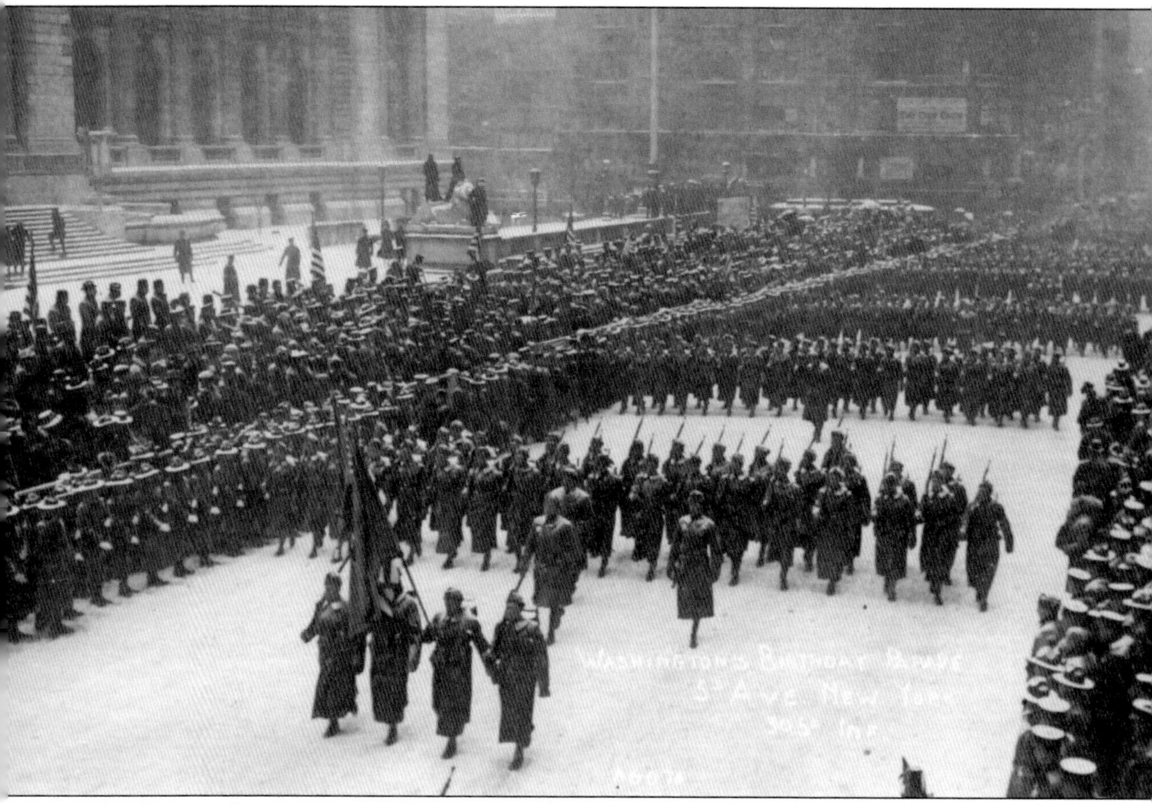

As the time for departure to Europe grew closer, it was decided to transport the 77th Division to New York City to march in a parade down Fifth Avenue on Washington's Birthday, February 22, 1918. The parade was followed by a military ball at the 71st Regiment Armory. The Division passes the New York Public Library at Forty-Second Street and Fifth Avenue. All discussion of the Army breaking up the 77th Division ended after this parade. When packing for overseas service, it was required that each division paint an identifying symbol on its baggage. The 77th Division chose the Statue of Liberty. After the Armistice, the Divisions were ordered to select and wear a Divisional insignia patch. The 77th chose as its emblem a golden Statue of Liberty on a blue field.

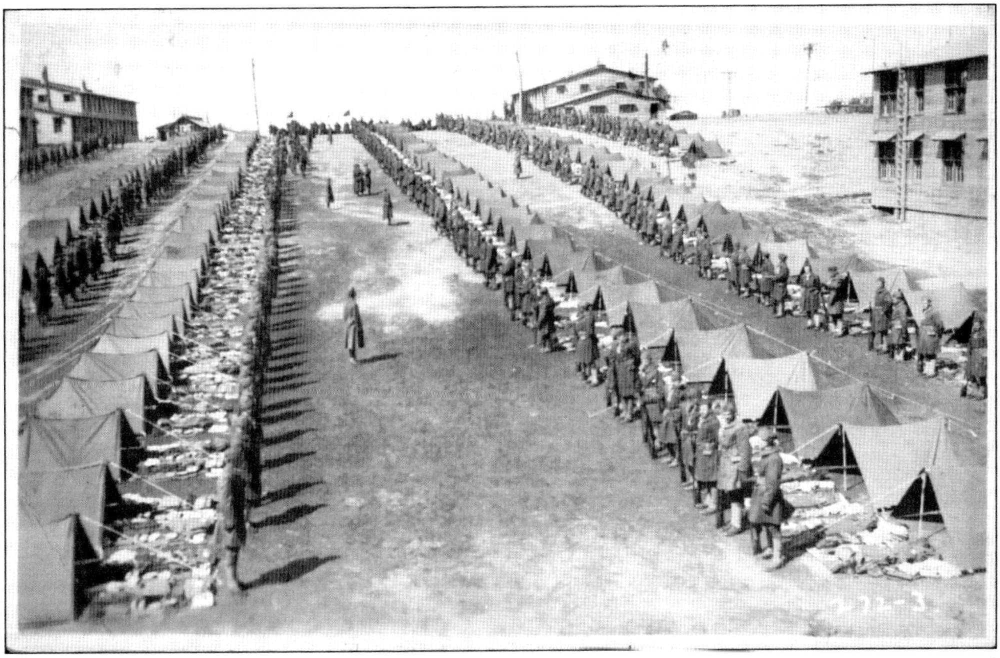

In March 1918, inspection of all issued equipment they would carry took place as troops prepared to depart for France. In his diary, Cpl. Oscar Ostrow of Brooklyn reported, "every man carried 100 rounds of ammo. Packs felt like pianos."

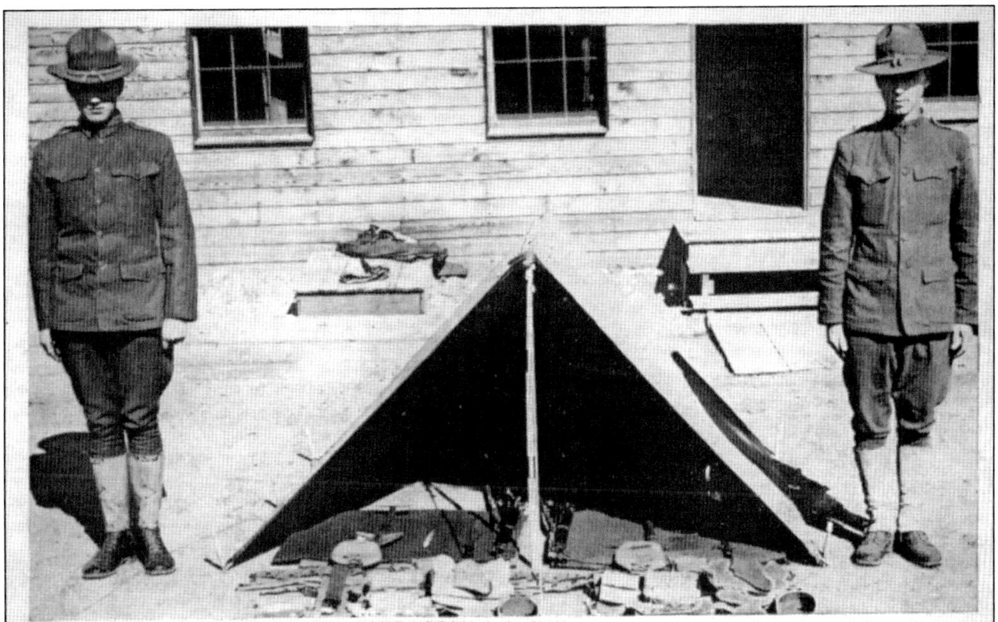

COMPLETE OUTFIT OF TWO MEN. CAMP UPTON, L. I., N. Y.

The soldiers described an "age of checking" that took place at Camp Upton before they left for France. These enlisted men have arranged all their issued equipment in the prescribed pattern and are ready for inspection. This included spare clothing, mess kit, containers for issued rations, and sealed metal cans containing hard bread.

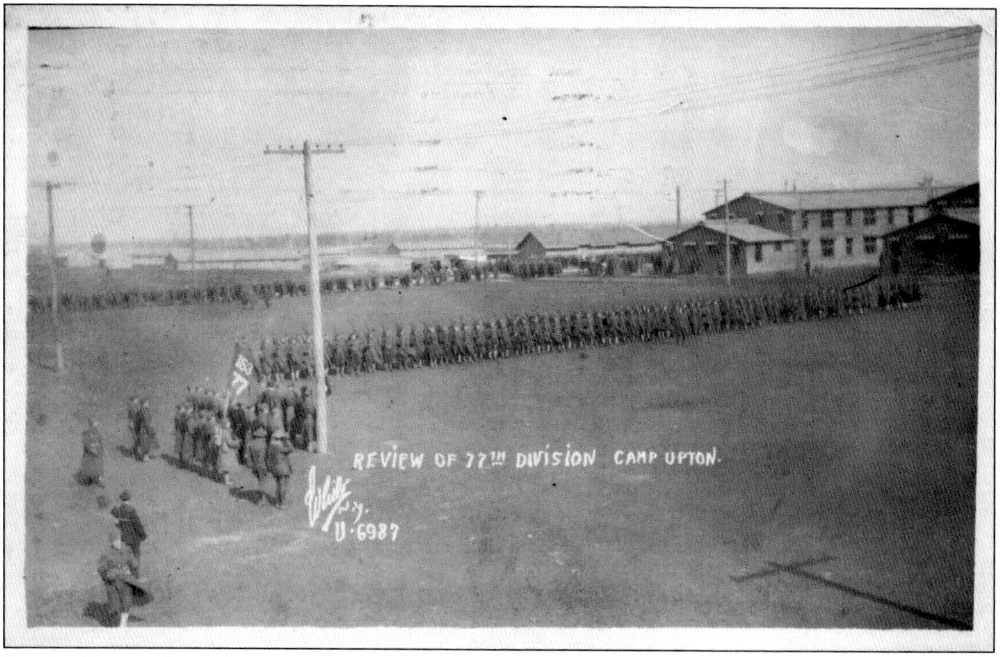

In March 1918, the 77th Division had completed training and would soon travel by truck and train to the embarkation camp at Camp Merritt in New Jersey and from there head to the port of Hoboken to board ships for France. The division would be under the charge of Gen. Evan M. Johnson. Due to poor health, General Bell was not able to accompany the troops he had trained.

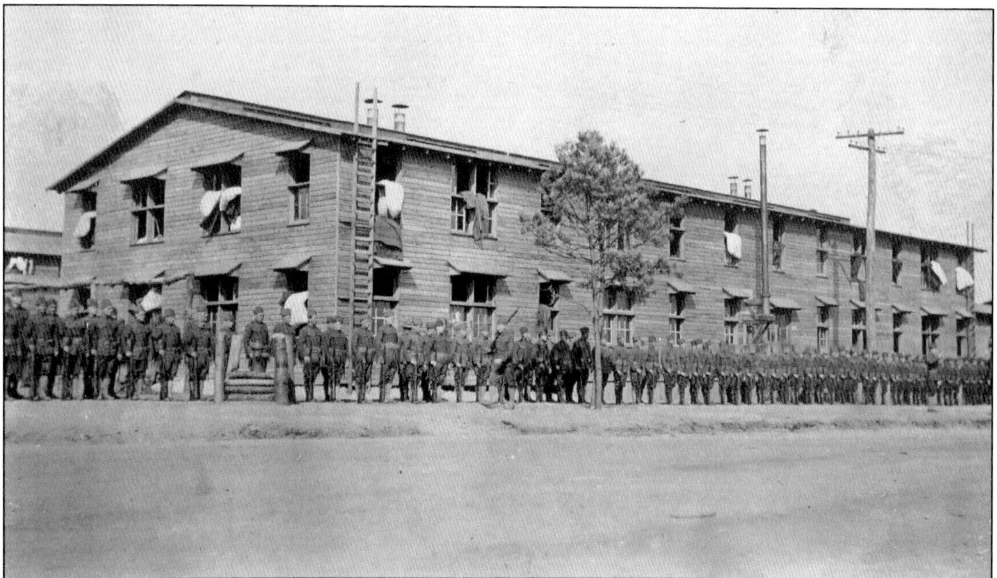

Co. I, 305 Inf. Camp Upton, L.I. "At Retreat."

"This is the company Jesse and Ralph were in at Camp Upton. Also their barracks. Jesse stands at extreme left 2nd one from end. Front row. Don't know where Ralph was in the line. Taken at 4pm April 1st 1918 as they were being discharged for the day." Although never mailed, the writer of this postcard provides an eyewitness account, as he most likely packed up and departed for France soon after this photograph was taken.

Three

Life at Camp Upton

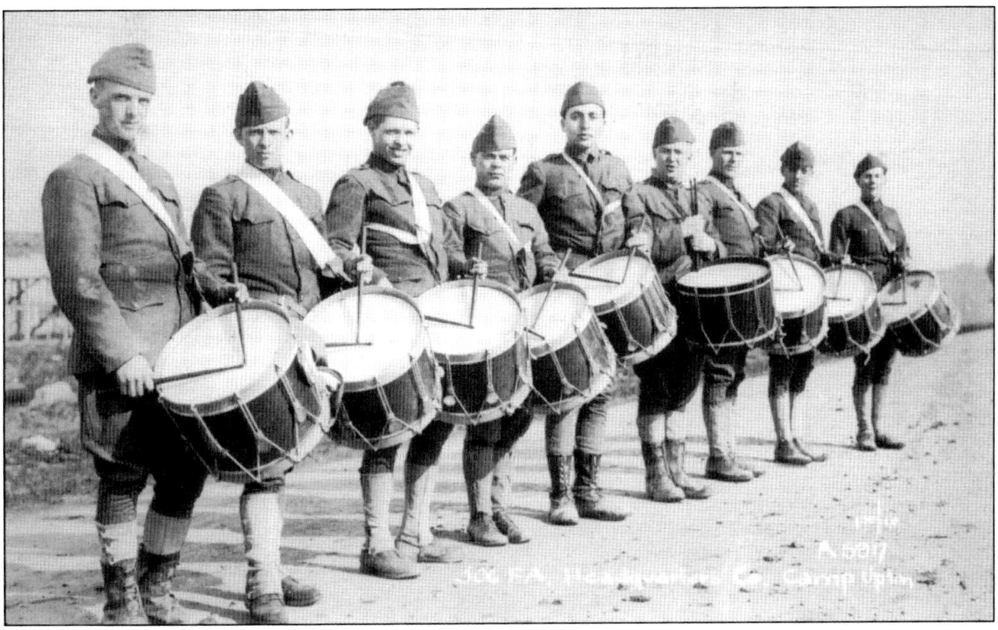

The 306th Field Artillery Headquarters Company drummers perform at General Bell's request. Composer John Philip Sousa revised an older song in 1917 as "The US Field Artillery March," featuring the vivid lyrics "Over hill, over dale, as we hit the dusty trail, Keep those caissons a rollin' along."

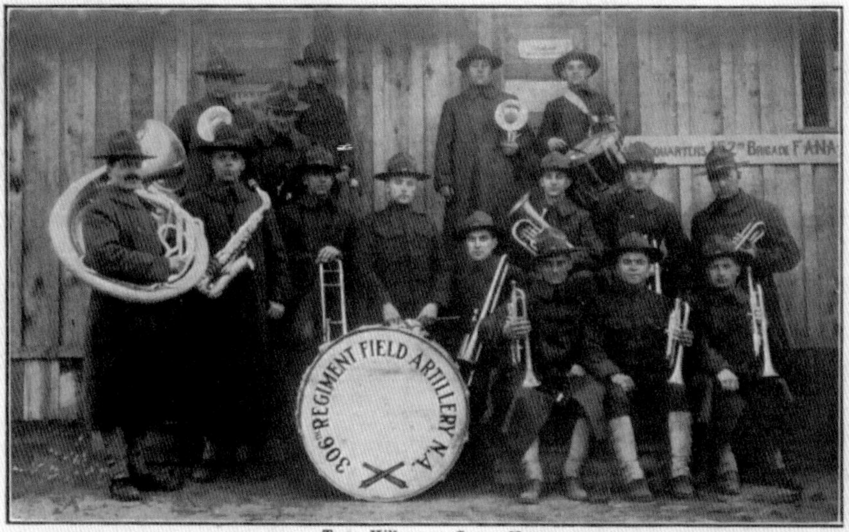

Tune Killers at Camp Upton.

Music was an important part of life at Camp Upton. Each regiment at Camp Upton was required to have a band. Many soldiers came to camp with musical instruments. The 306th Field Artillery Regiment band was just one of many at Camp Upton.

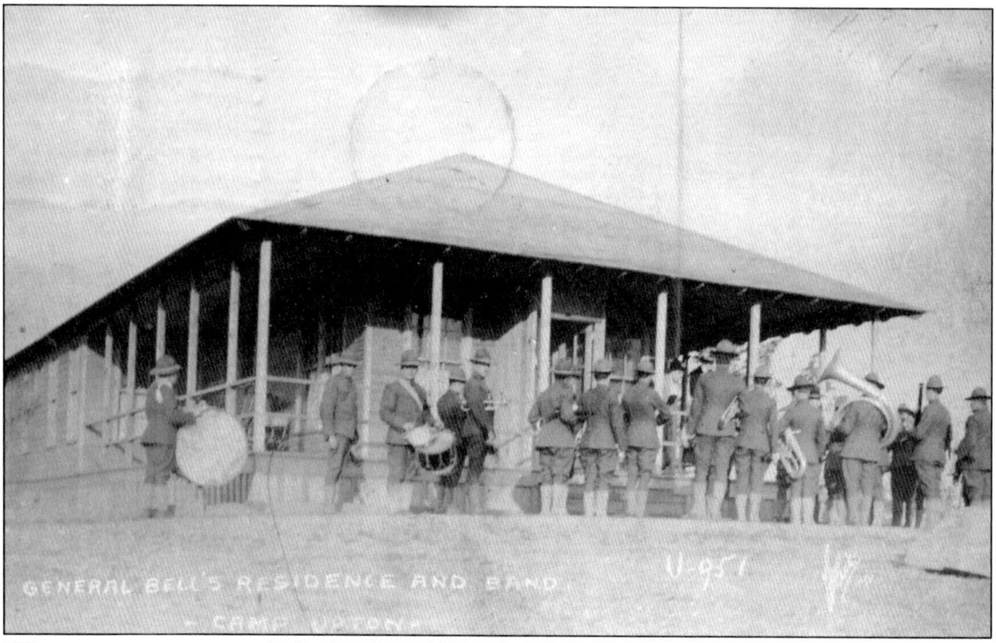

When new recruits arrived, the 304th Field Artillery band would be ordered to meet them at the station as their train pulled in. Major General Bell sent for the band to entertain distinguished guests at the camp.

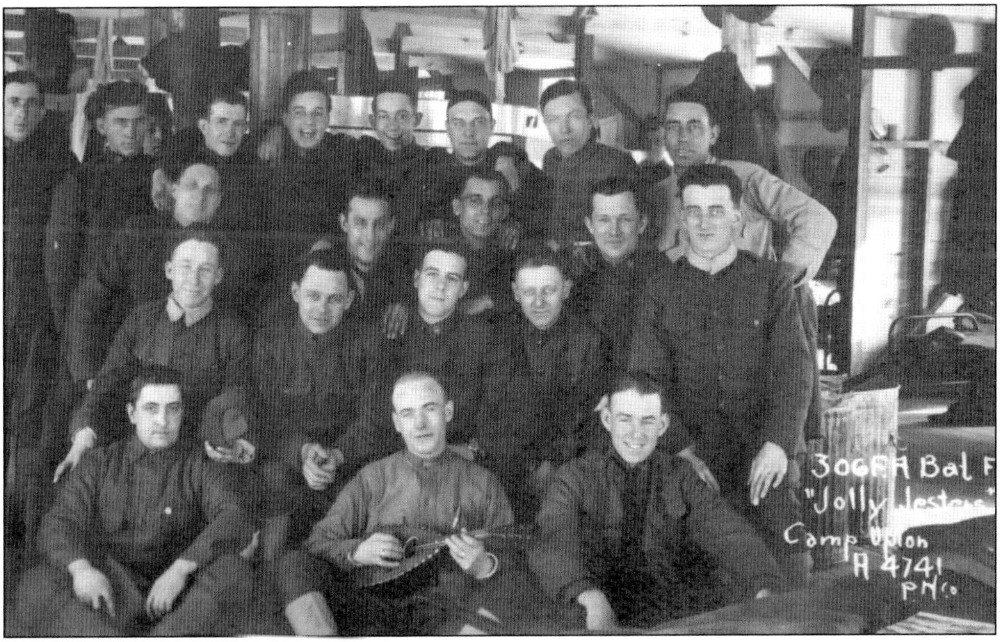

General Bell encouraged singing and music to lift the spirits of the men. The Jolly Jesters of Battery F, 306th Field Artillery, are shown in their barracks. Music and singing were an important part of army life, and General Bell used them to boost morale.

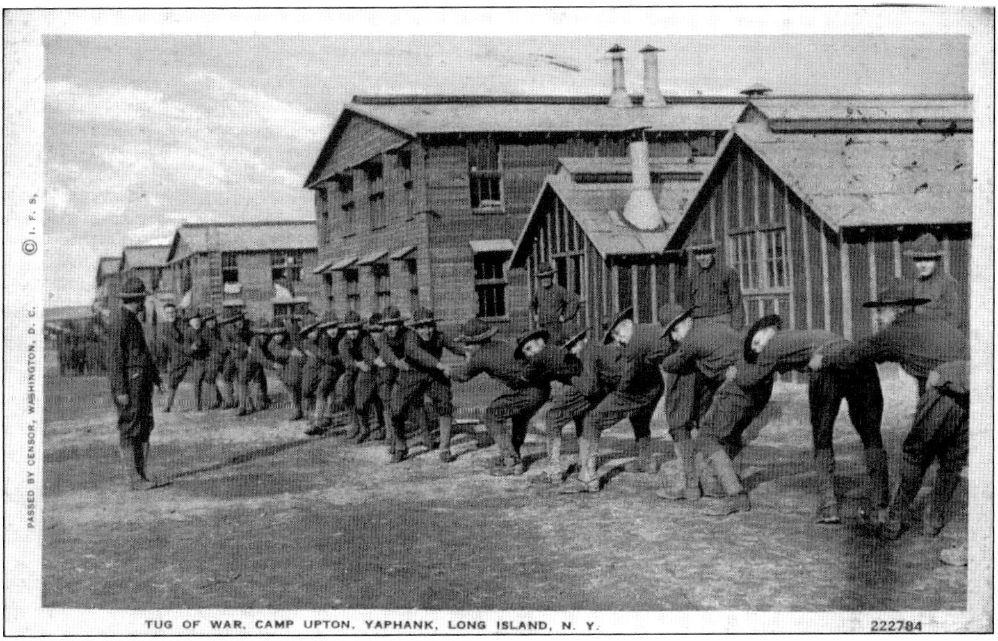

Tug-of-war was a popular sport that combined physical conditioning and directed cooperative effort. It was part of the Olympic Games from 1900 until 1920 but has not been included since.

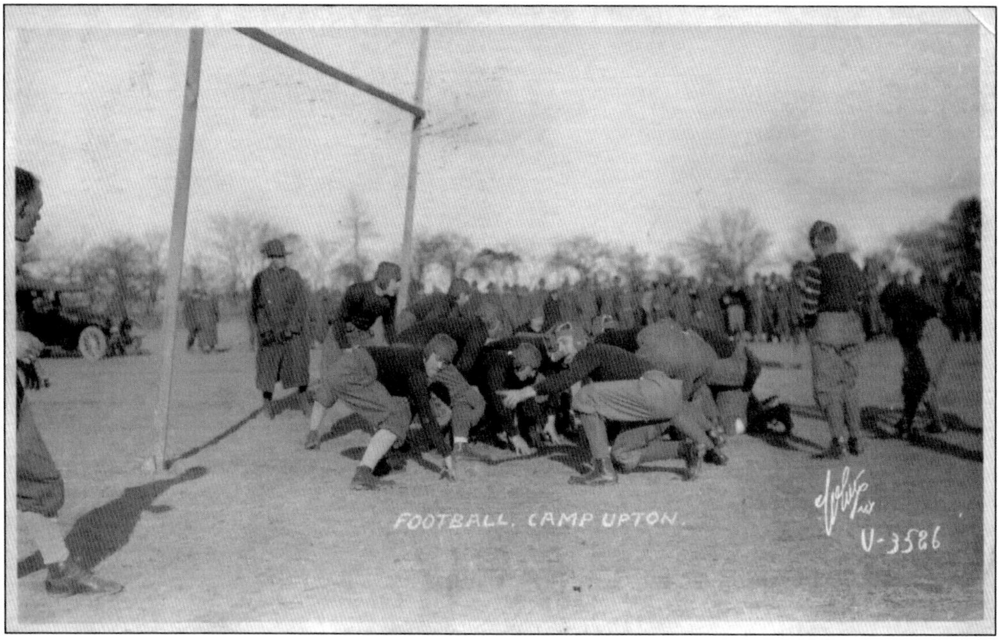

Frank Glick, captain of the 1915 Princeton University football team, was engaged as a civilian aid to General Bell to take charge of athletics at the camp. Every company had a team in a variety of sports. According to their regimental history, the 302nd Engineers had the best football team.

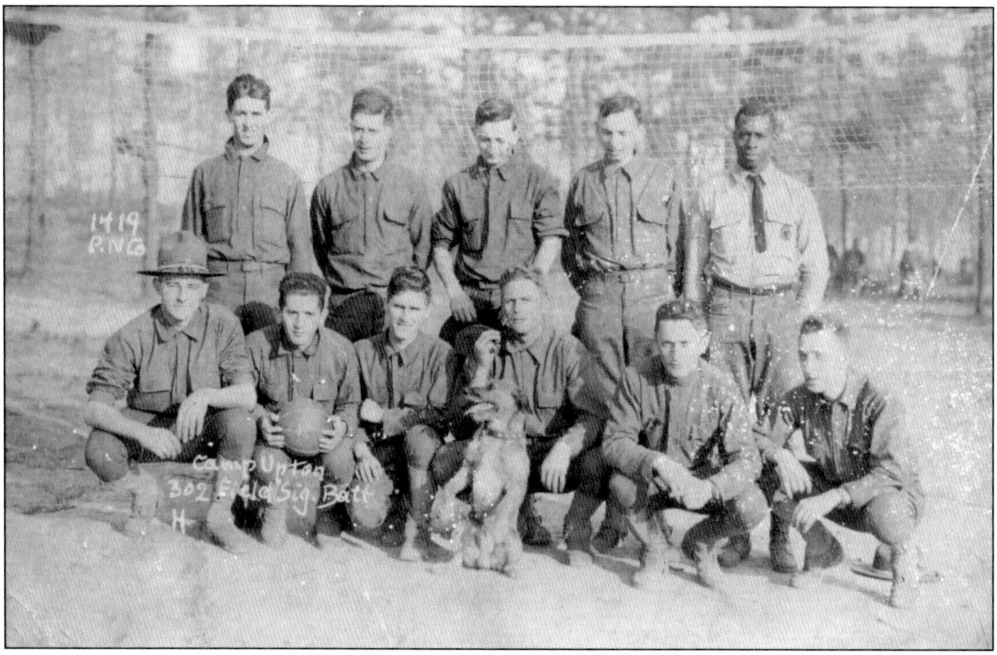

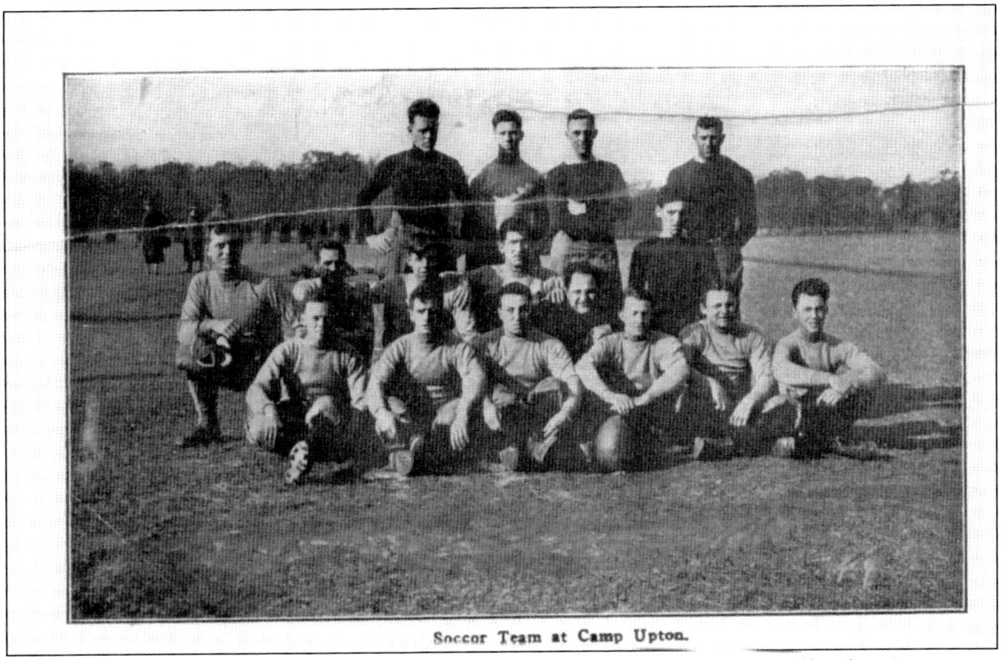

Although the caption on this postcard claims this is a "soccor" team, it is actually what Americans call a football team. Note the ball, helmet, and cool-weather uniforms.

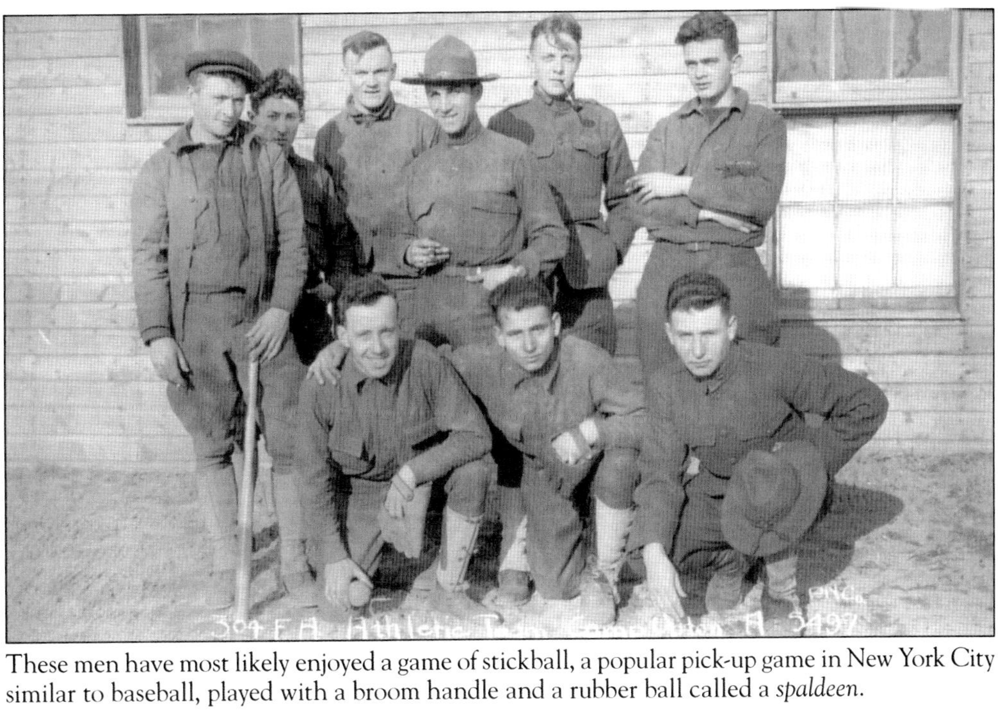

These men have most likely enjoyed a game of stickball, a popular pick-up game in New York City similar to baseball, played with a broom handle and a rubber ball called a *spaldeen*.

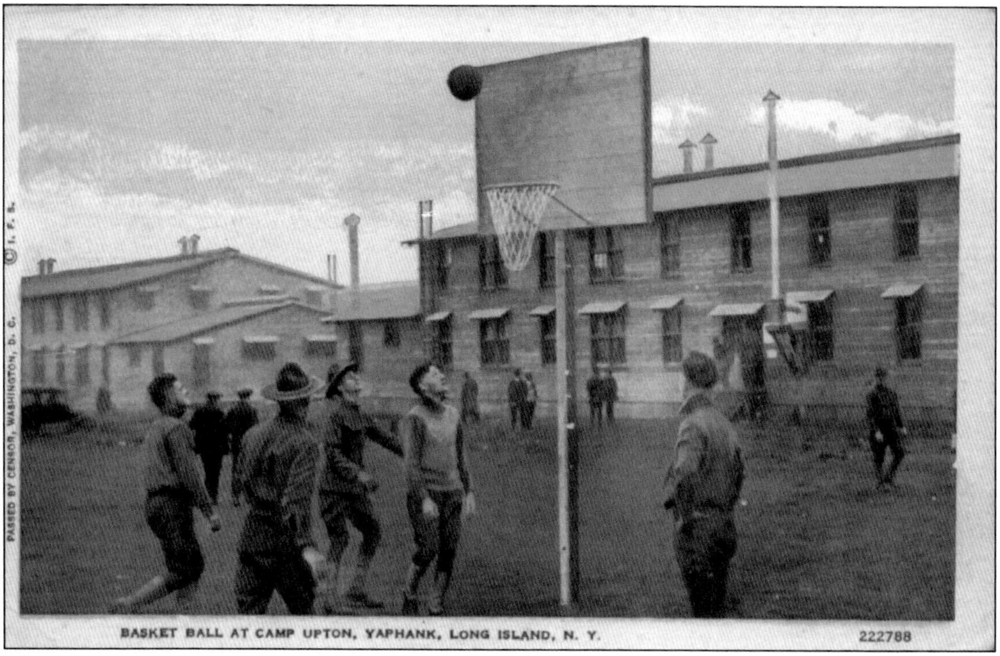

The Depot Brigade constructed basketball courts for the men. Teams from each regiment would form and play against each other and against teams from other camps. Teams from Camp Devens in Massachusetts visited often.

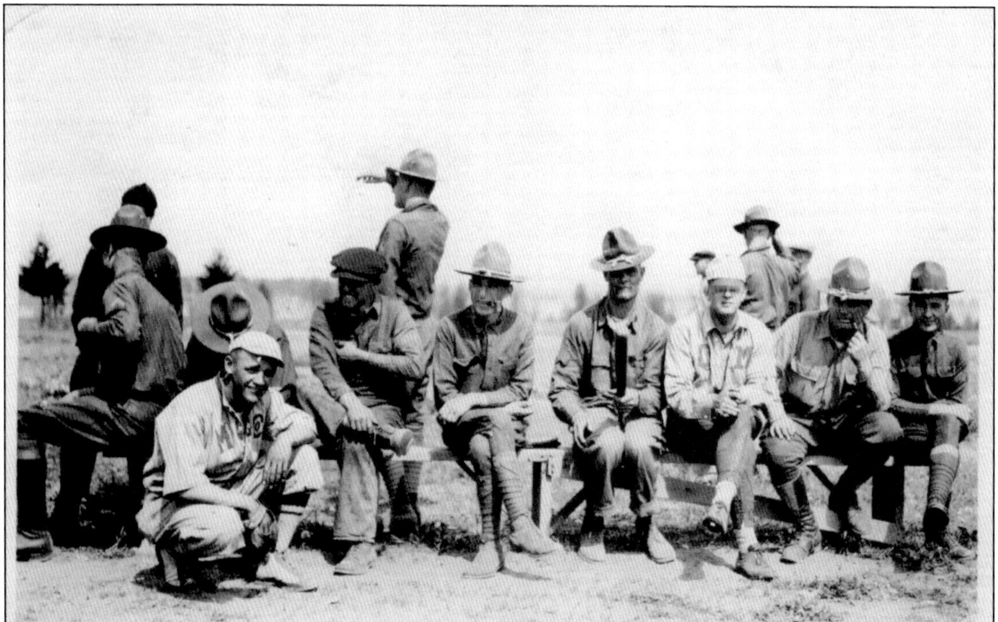

Ample playing fields were established between the barracks. Eddie Grant, ex–New York Giants third baseman and shortstop, organized baseball teams.

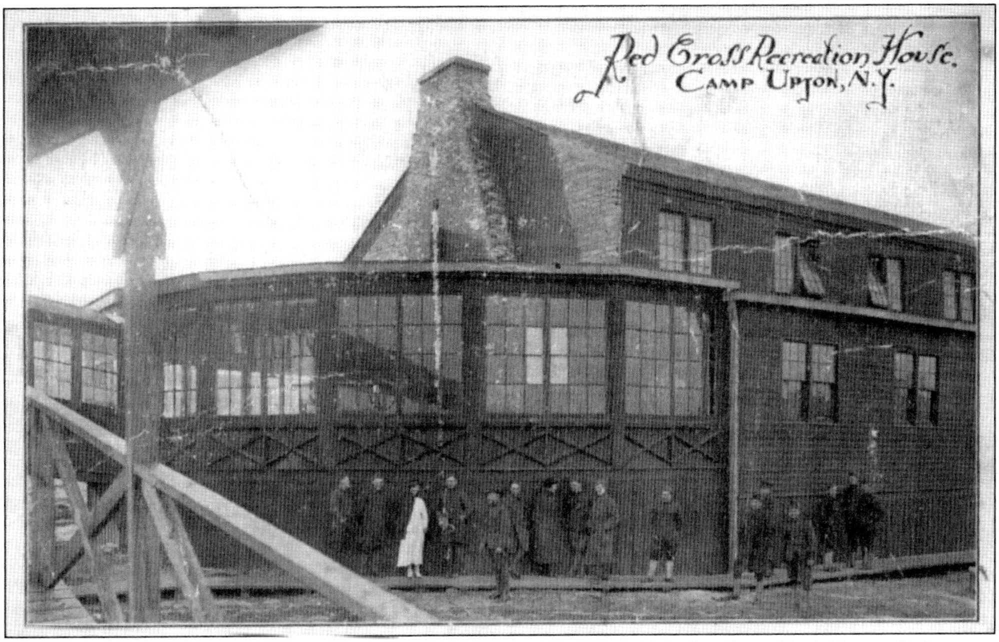

Service organizations such as the YMCA, YWCA, Knights of Columbus, Red Cross, Salvation Army, and the Jewish Welfare Board built structures at the camp to provide respite and recreation for the soldiers. Each site encouraged wholesome activities such as reading, writing, and games.

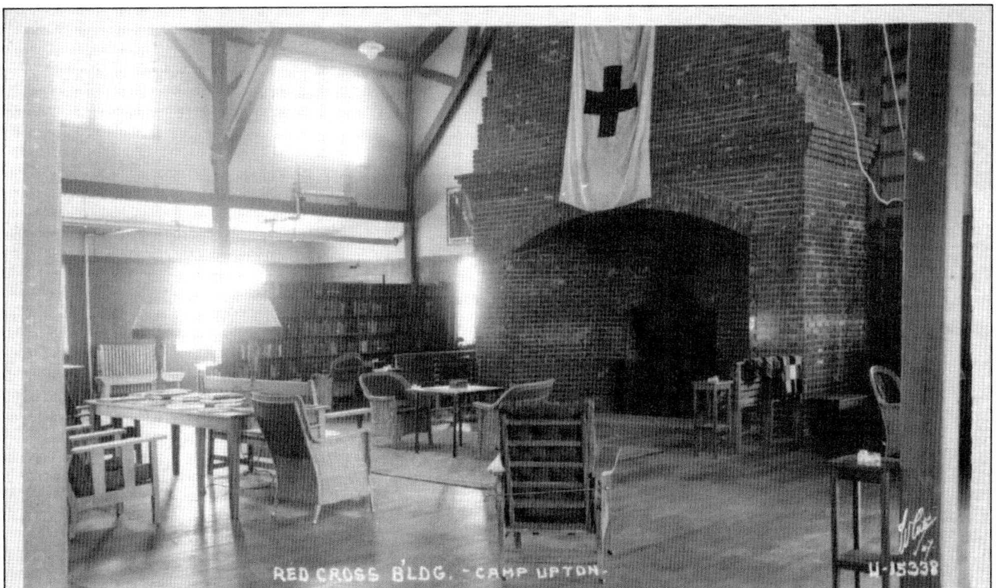

The Red Cross, established in 1882 by Clara Barton, grew rapidly in 1918 as it provided services to the American armed forces and their families. Local chapters were established to support efforts at the camp. Local women volunteered, including Hope Fullerton of Huntington, the daughter of Long Island Rail Road special agent and photographer Hal B. Fullerton. She often spoke about how many supplies they readied and packed for shipping overseas: "they made bandages by the ton."

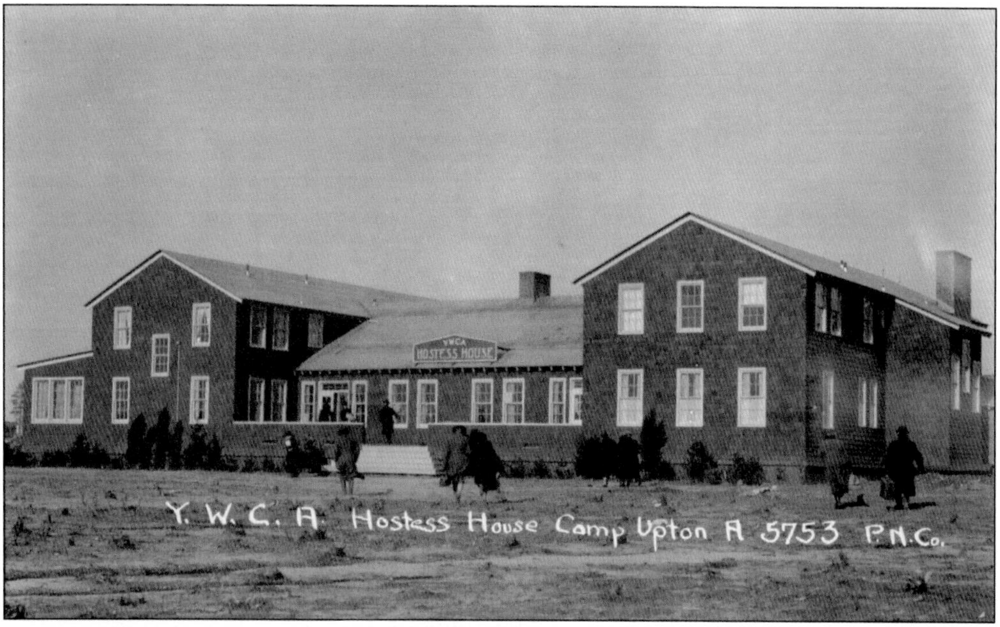

At Camp Upton, the Young Women's Christian Association (YWCA) built several hostess houses where soldiers could meet with family and friends. The YWCA was the first organization to send professional workers overseas to provide administrative leadership and support to the US armed forces in 1918.

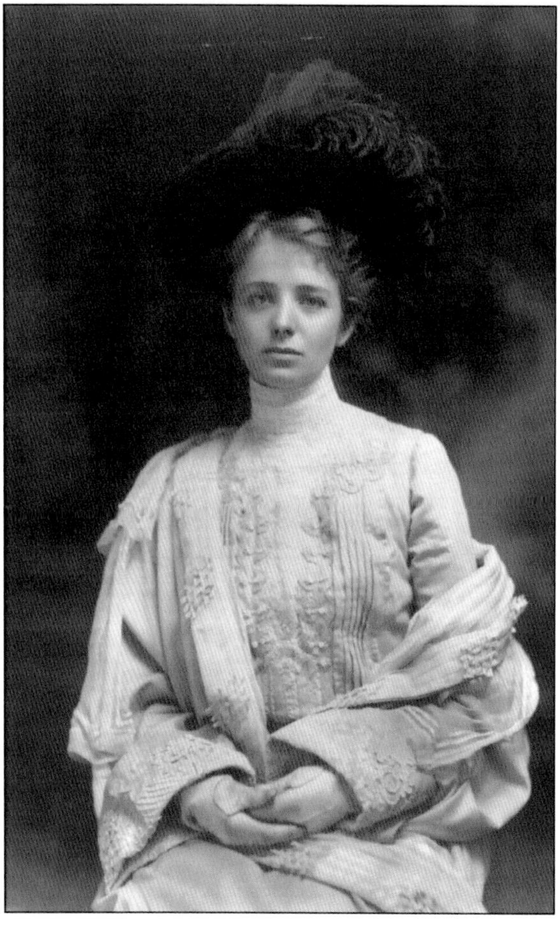

Noted stage actress Maude Adams (1872–1953) is seen here in 1902. She was the first actress to play Peter Pan on Broadway. She spent summers in nearby Lake Ronkonkoma and volunteered at Camp Upton in 1918 while taking a break from performing. She stayed at the YWCA Hostess House, served meals, folded napkins, arranged flowers, and visited soldiers in the base hospital. (Courtesy of the Library of Congress.)

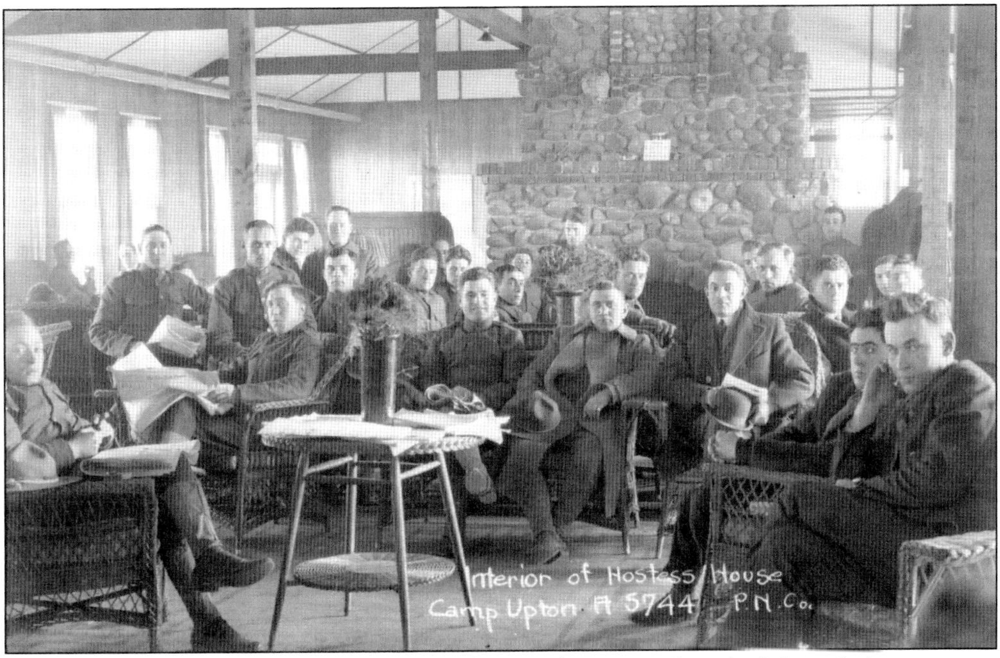

Comfortable YWCA Hostess Houses for the officers and men were furnished through the generosity of a group of Long Island women. Soldiers would write letters, play games, and read.

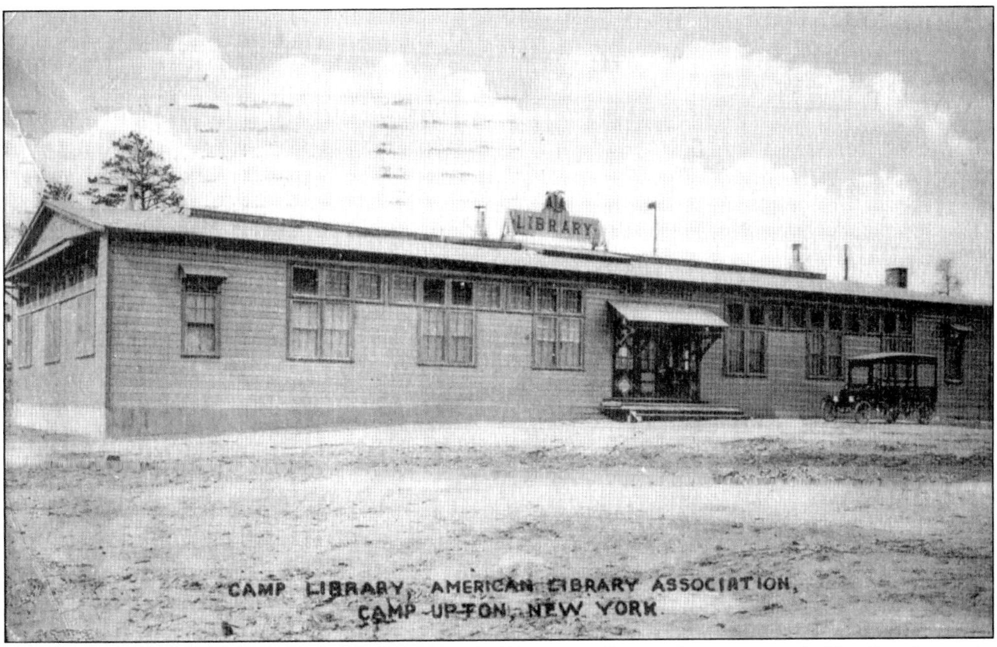

The American Library Association maintained a library at Camp Upton, provided by the Carnegie Library Fund, with a stock of 8,000 books, newspapers, and magazines, eventually circulating 186,000 items. There were also 22 library branches throughout the camp. Camp Upton's regular newspaper, *Trench and Camp*, was published from October 8, 1917, until August 26, 1919.

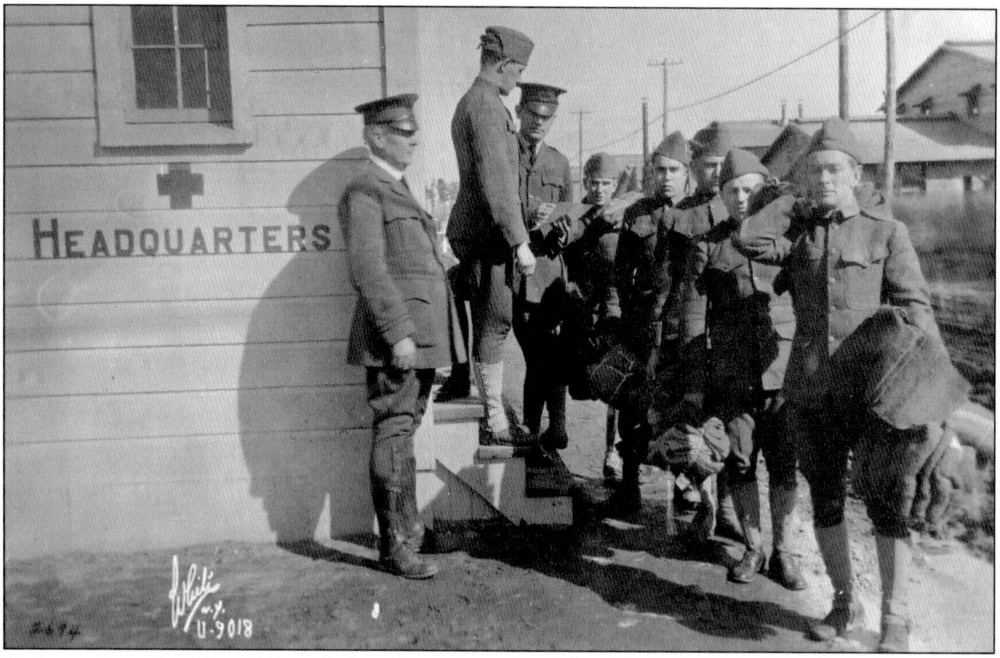

Packages from family and friends from home were greatly appreciated by the soldiers during their training. Women organized "knitting bees" to make garments and socks for the soldiers. Soon, many would be boarding ships for Europe's battlefields. The crossing took 8 to 10 days. (Courtesy of the Library of Congress.)

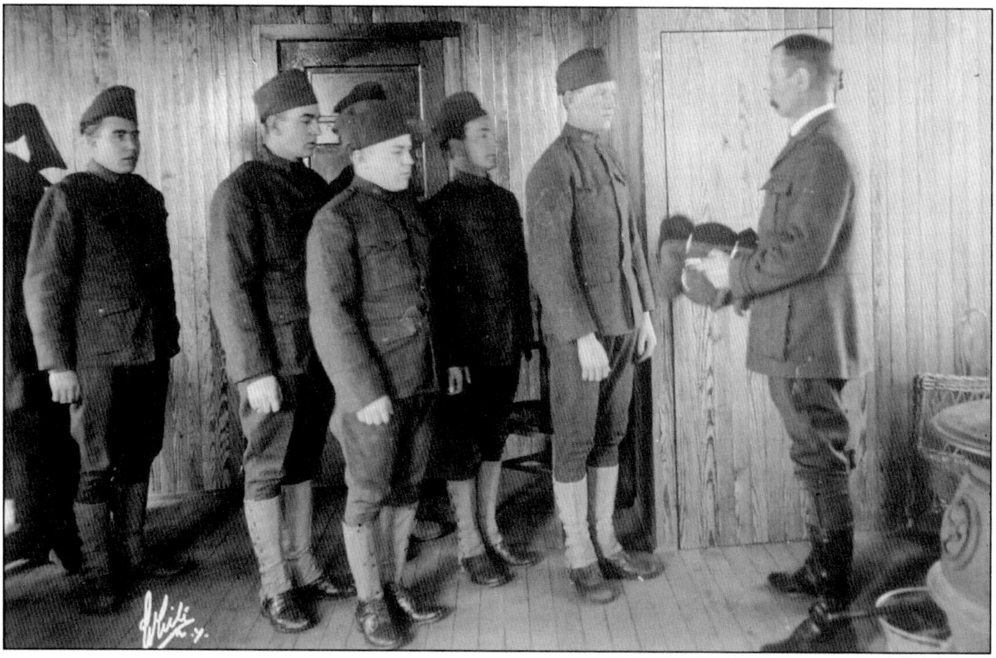

In July 1918, about 3,000 knitters worked for three days in New York's Central Park to knit wool socks, sweaters, and mufflers for servicemen. The American Red Cross then delivered them to soldiers at the camp. Officer Topham is shown in his Red Cross uniform. (Courtesy of the Library of Congress.)

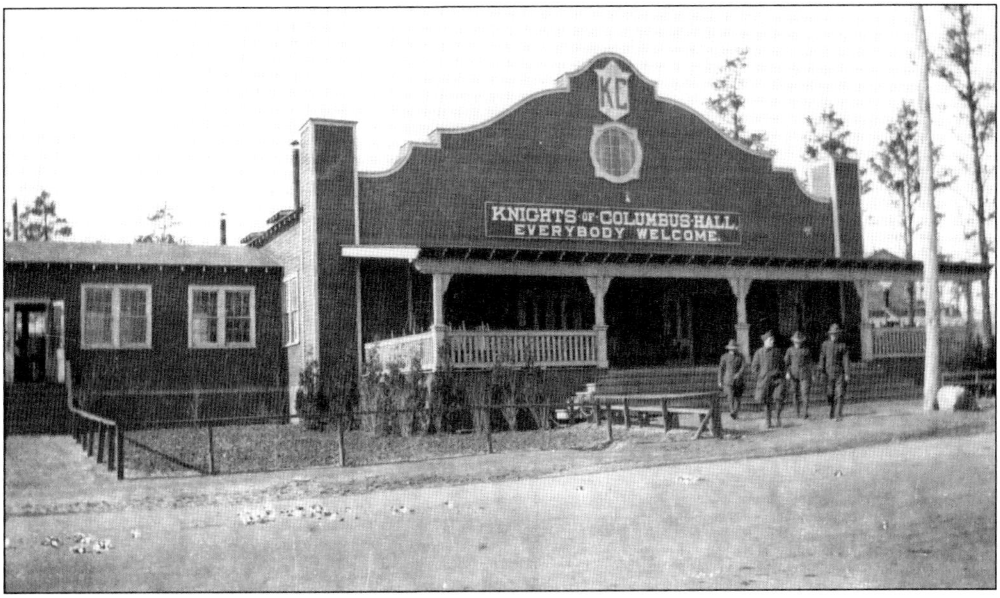

Many large events were held at the Knights of Columbus building, including movies, concerts, dances, and boxing matches. Services and facilities were open to all, regardless of race or creed. Their motto was "Everybody Welcome, Everything Free." The building burned down in March 1918, possibly due to crossed wires.

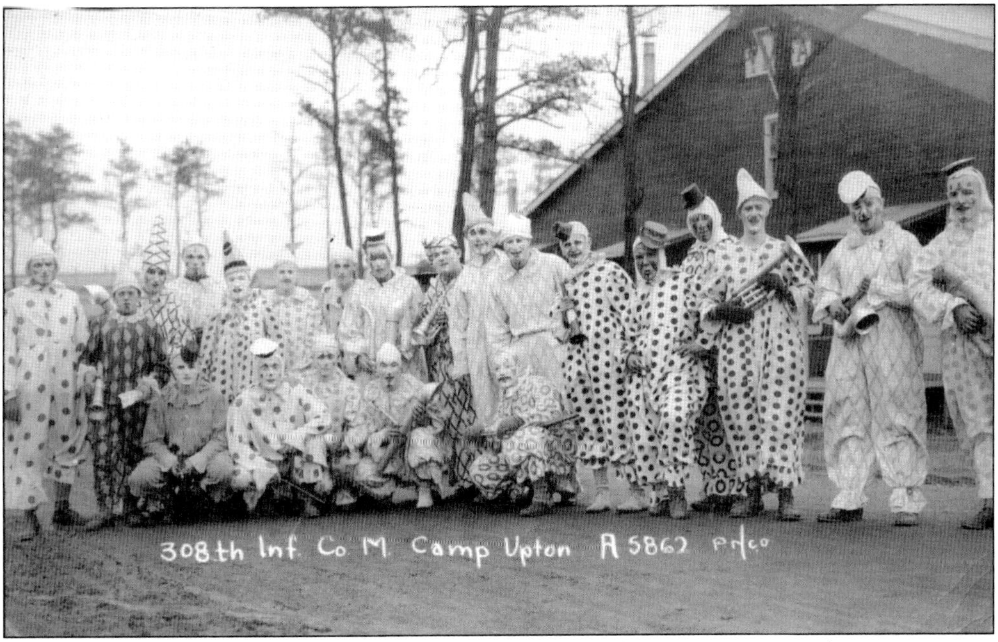

The 308th Regiment held a circus on Lincoln's birthday in February 1918 to provide entertainment for the soldiers. Members of the 308th Infantry dressed as clowns and led a parade, which also featured animals and daredevils.

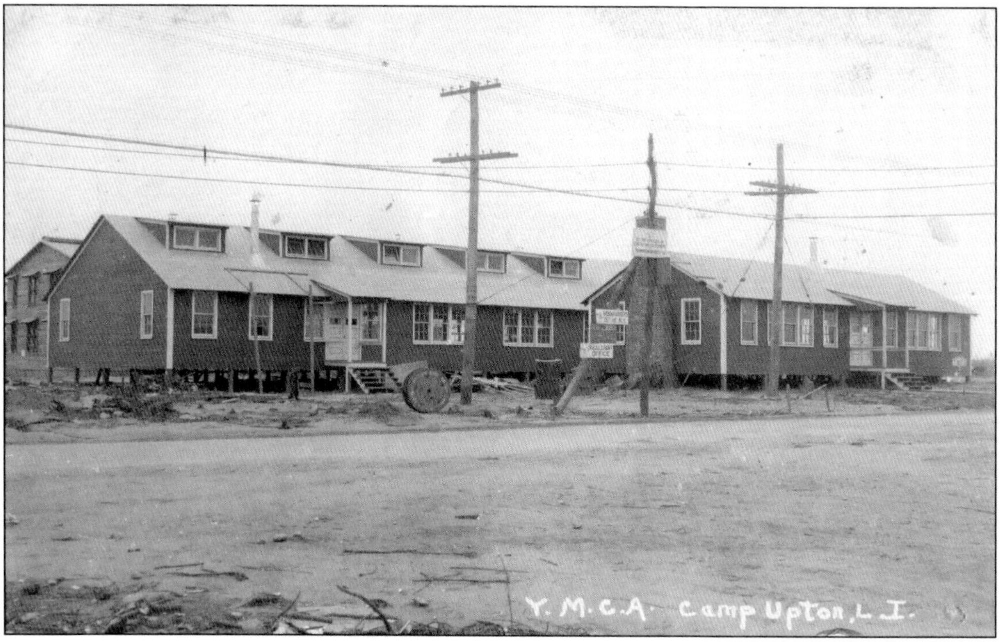

The Young Men's Christian Association (YMCA) was founded in London in 1844, and the first YMCA in America opened in Boston in 1851. The organization provided morale and welfare services for the military and operated YMCA huts for recreation and religious services. Each regiment had its own YMCA hut.

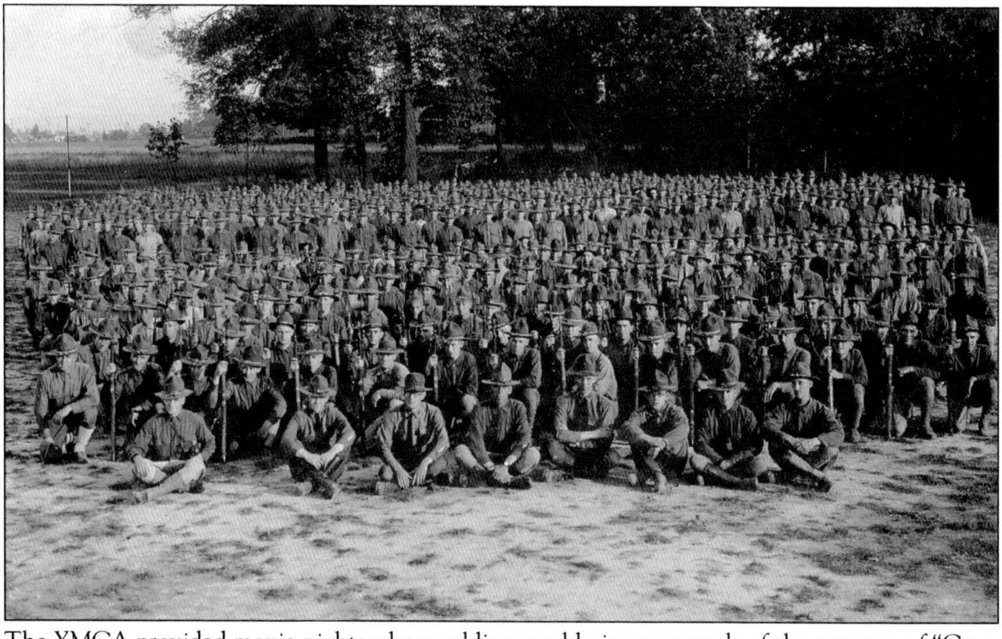
The YMCA provided movie nights where soldiers could view newsreels of the progress of "Gen. Pershing's Crusaders" in Europe. Movies were screened outside to diminish the spread of flu in September 1918.

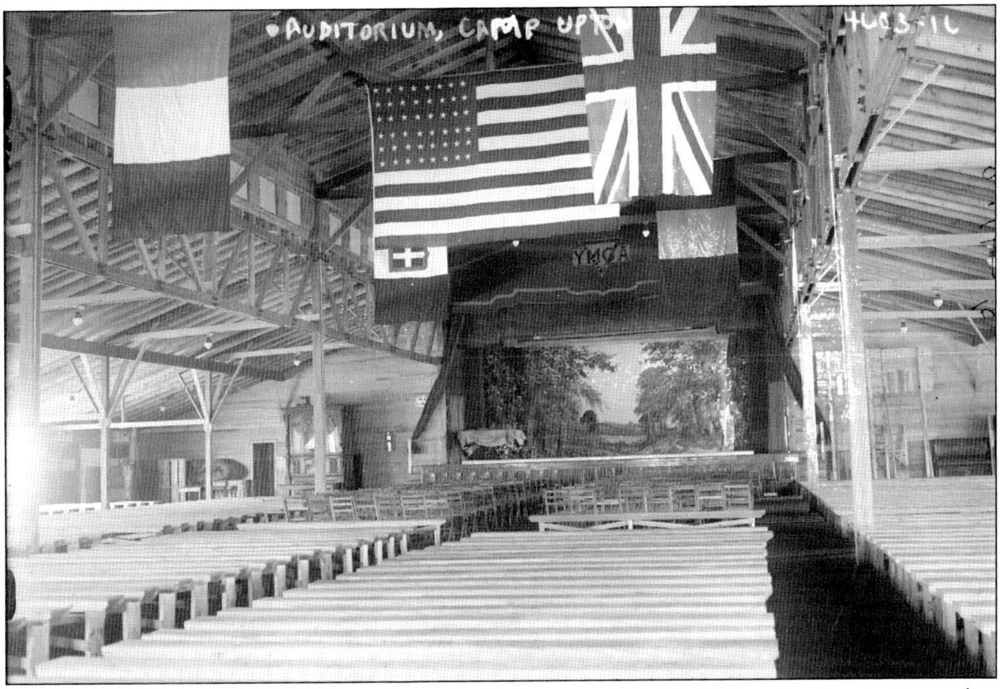

Former president William H. Taft spoke at the dedication of the YMCA auditorium in December 1917. The New York Symphony played in the YMCA auditorium, as did the New York Philharmonic. The auditorium could seat 3,000.

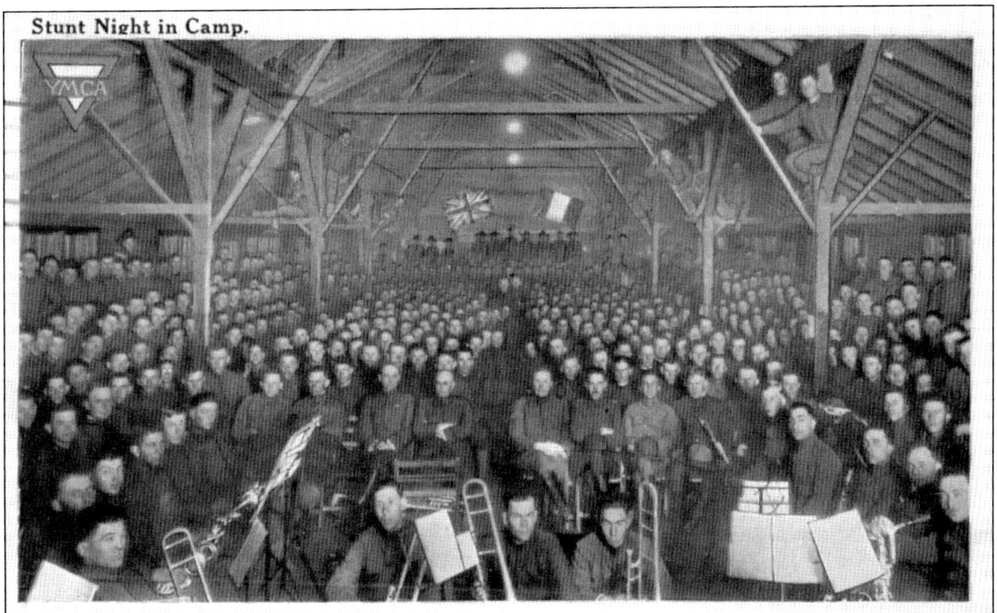

YMCA facilities around the country were integrated, and Camp Upton was no exception. Clashes began in April 1918. After white soldiers from Texas objected to two black soldiers writing letters in the YMCA, General Bell instructed his officers that they were responsible for the acts of their men and told them that all recreational facilities, including the YMCA buildings and theatres, were open to white and black troops.

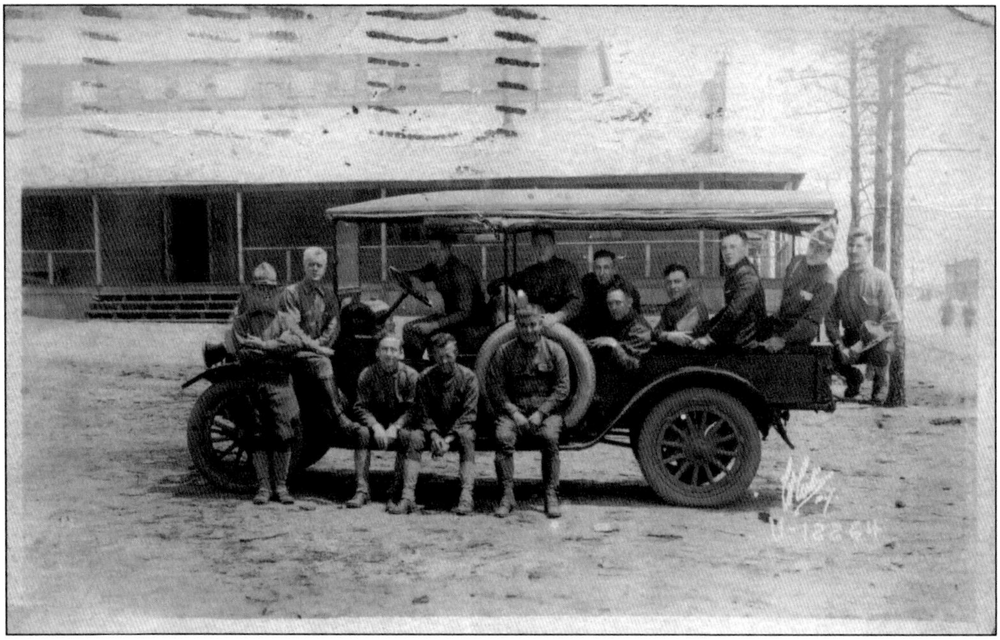

Local men such as Albert Bayles and Charles Miller offered a popular service picking up soldiers in their trucks and bringing them to Patchogue or into the city. Soldiers went into New York City on 48- and 36-hour passes that had to be used within 10 days. Contracting venereal disease prior to being sent to France was a concern, as was drunkenness. Alcohol was prohibited for sale within five miles of the camp.

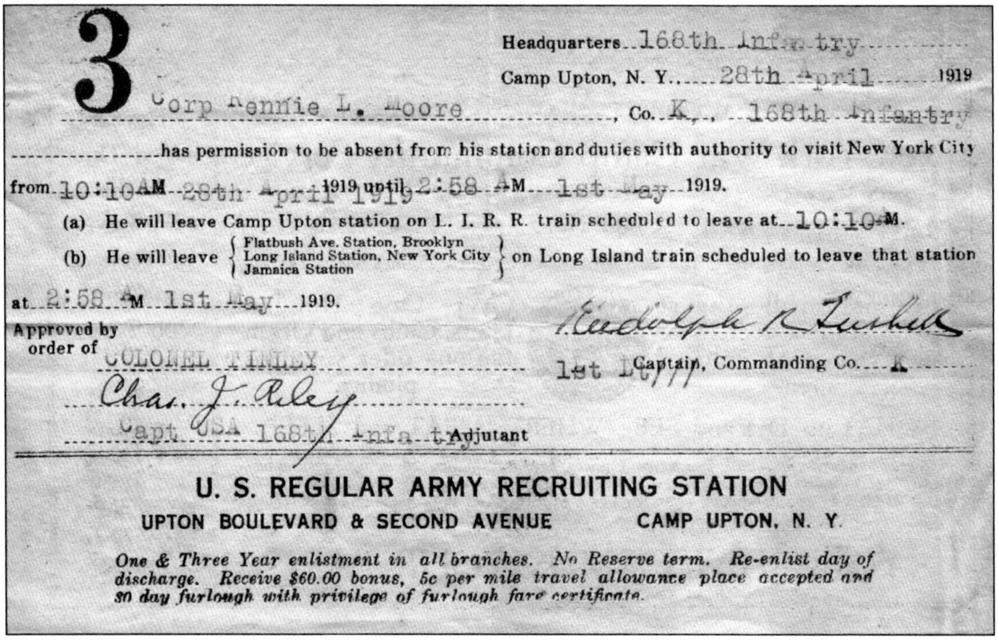

3

Headquarters 168th Infantry
Camp Upton, N. Y. 28th April 1919

Corp Rennie L. Moore, Co. K, 168th Infantry

has permission to be absent from his station and duties with authority to visit New York City

from 10:10 AM 28th April 1919 until 2:58 AM 1st May 1919.

(a) He will leave Camp Upton station on L. I. R. R. train scheduled to leave at 10:10 AM.

(b) He will leave { Flatbush Ave. Station, Brooklyn / Long Island Station, New York City / Jamaica Station } on Long Island train scheduled to leave that station

at 2:58 AM 1st May 1919.

Approved by order of COLONEL TINLEY

Chas. J. Riley
Capt USA 168th Infantry Adjutant

Rudolph R. Tushell
1st Lt, Captain, Commanding Co. K

U. S. REGULAR ARMY RECRUITING STATION

UPTON BOULEVARD & SECOND AVENUE CAMP UPTON, N. Y.

One & Three Year enlistment in all branches. No Reserve term. Re-enlist day of discharge. Receive $60.00 bonus, 5c per mile travel allowance place accepted and 30 day furlough with privilege of furlough fare certificate.

Cpl. Rennie L. Moore, 168th Infantry, Company K, was granted a three-day pass to New York City on April 28, 1919. Camp policy encouraged soldiers to "have a good time in the Big Town but get back to Camp on time." (Courtesy of the Suffolk County Historical Society.)

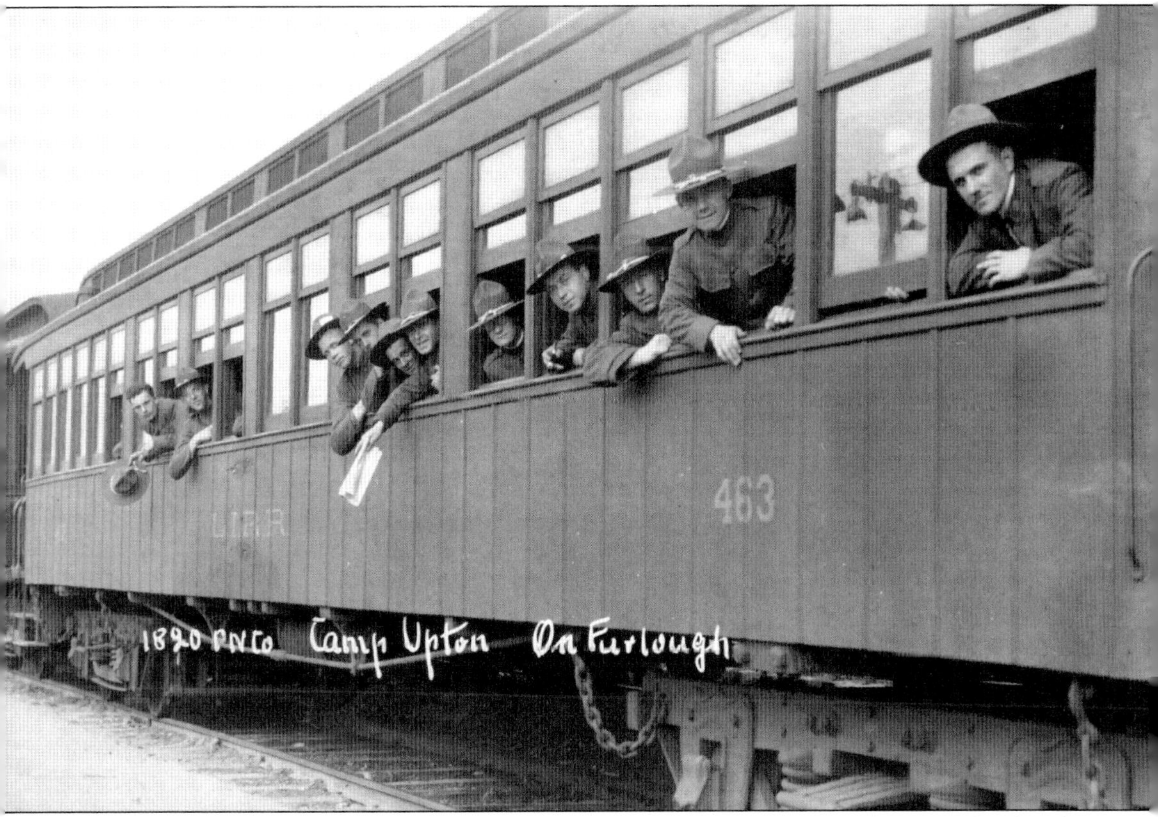

Many of the soldiers were from the New York City area and were able to board a Long Island Rail Road train at the camp for $1.30 round trip and return home on a pass to Brooklyn, Queens, and the Bronx for the weekend. The last train from New York was at 2:59 a.m., and it arrived in time for reveille on Sunday morning.

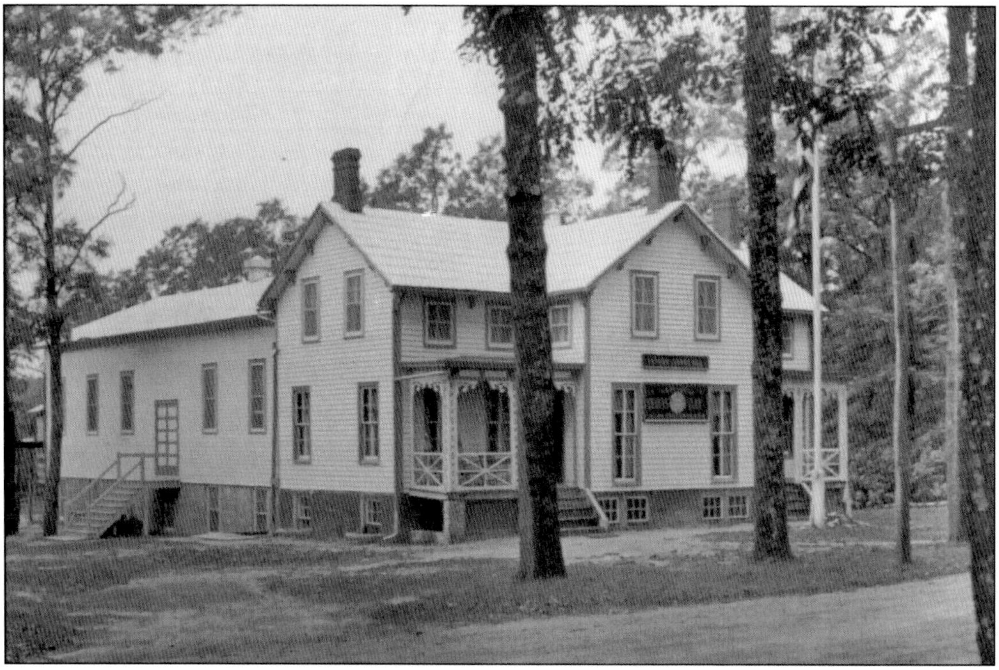

Yaphank's Grange Hall, built in 1913, attracted soldiers for an evening's entertainment away from the camp. It still stands on Main Street, converted to a bank many years ago.

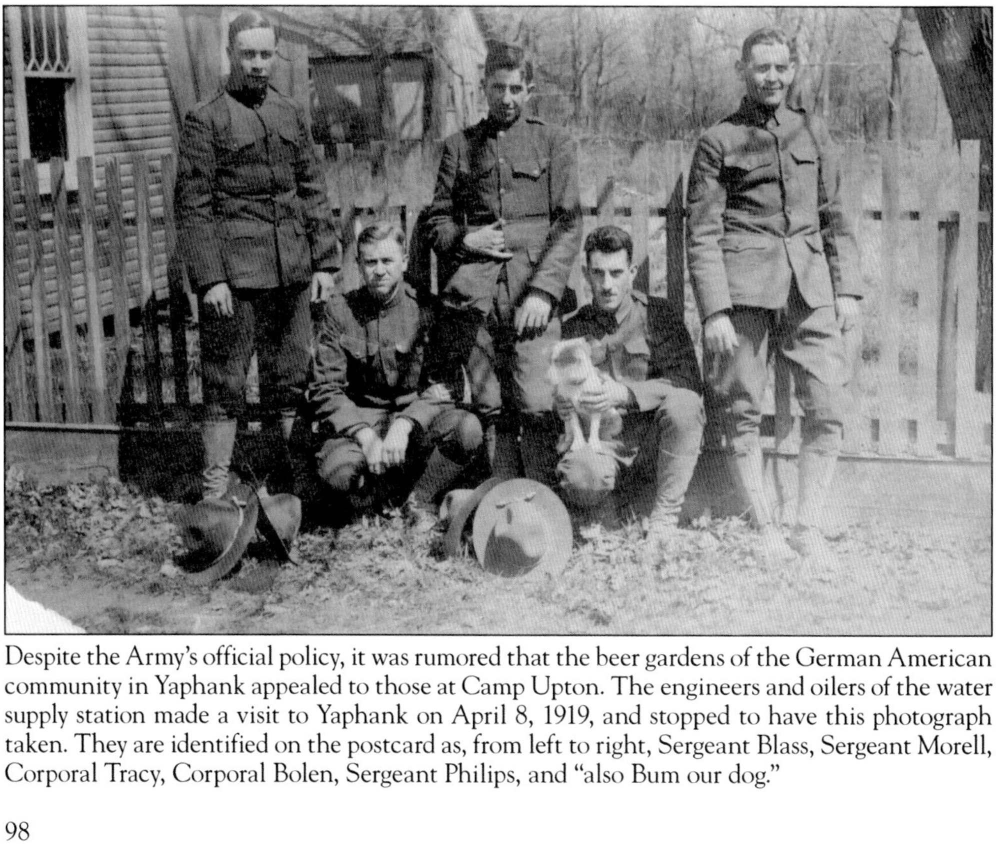

Despite the Army's official policy, it was rumored that the beer gardens of the German American community in Yaphank appealed to those at Camp Upton. The engineers and oilers of the water supply station made a visit to Yaphank on April 8, 1919, and stopped to have this photograph taken. They are identified on the postcard as, from left to right, Sergeant Blass, Sergeant Morell, Corporal Tracy, Corporal Bolen, Sergeant Philips, and "also Bum our dog."

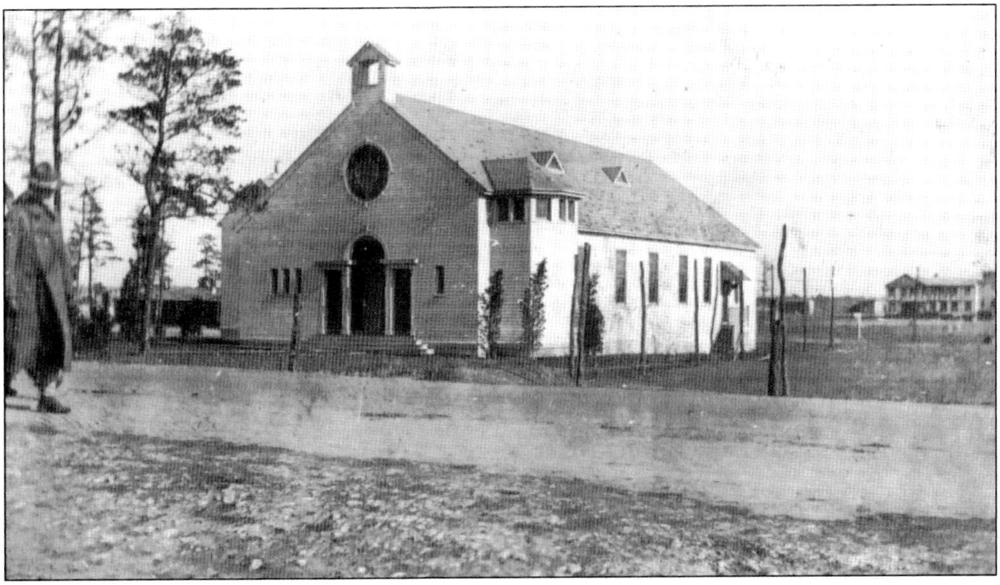

The camp chapel was where Protestant, Catholic, Jewish, and Christian Science services were held. The chapel cost $34,000 to build.

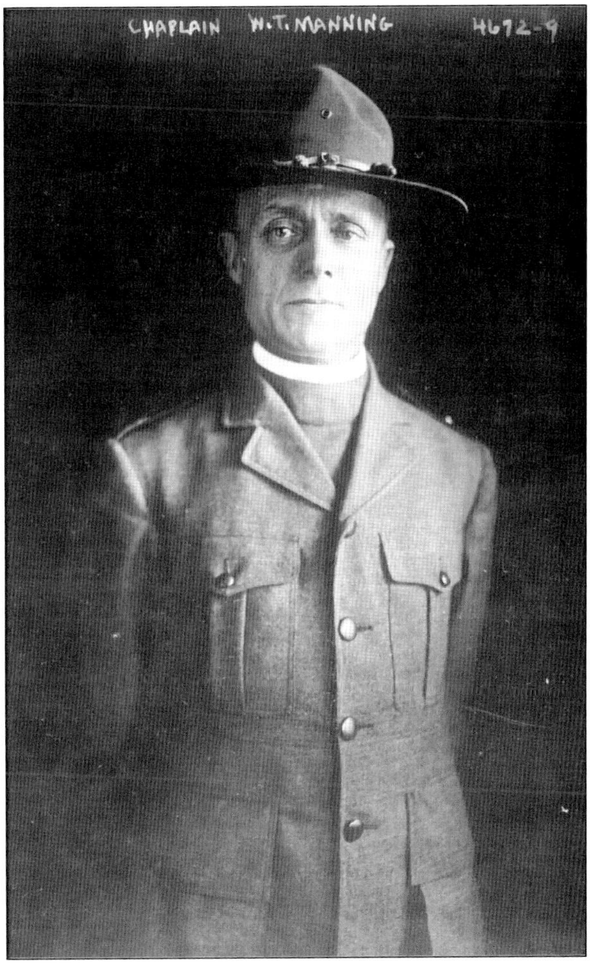

This photograph shows Rev. William Thomas Manning (1866–1949), rector of Trinity Church in New York and chaplain of the 302nd Engineers at Camp Upton, later Episcopal bishop of New York (1921–1946). (Courtesy of the Library of Congress.)

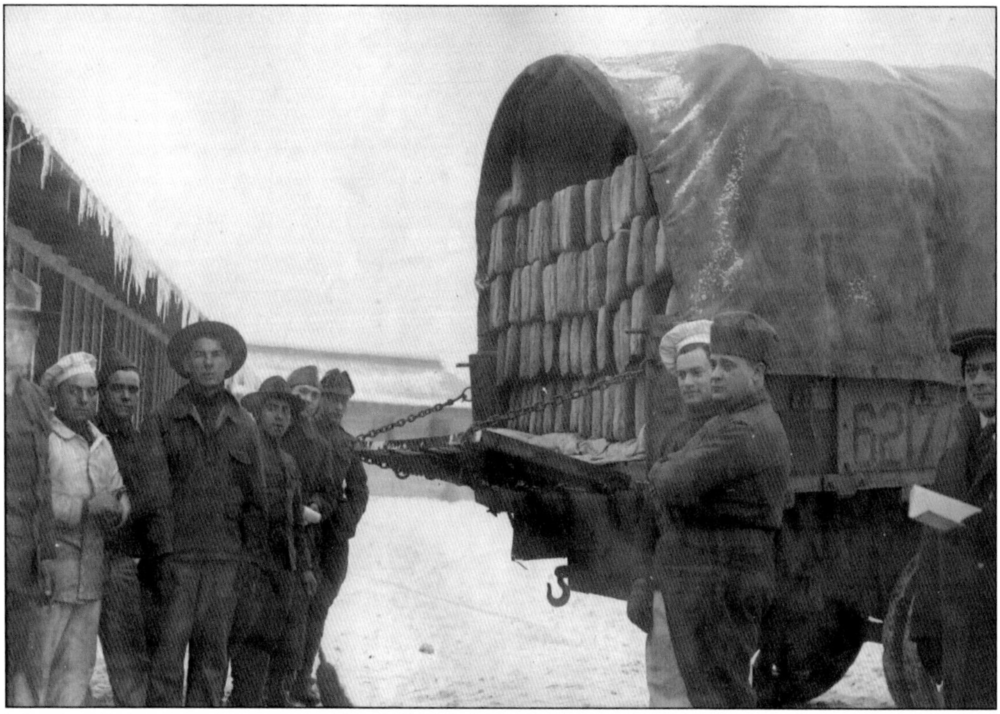

The field bakery was a warm and busy place in the frigid winter of 1918 at Camp Upton, operating 24 hours a day. The ovens produced 300 loaves of bread an hour. The loaves were held for a day, as they were considered "too new," then delivered to the mess halls by truck.

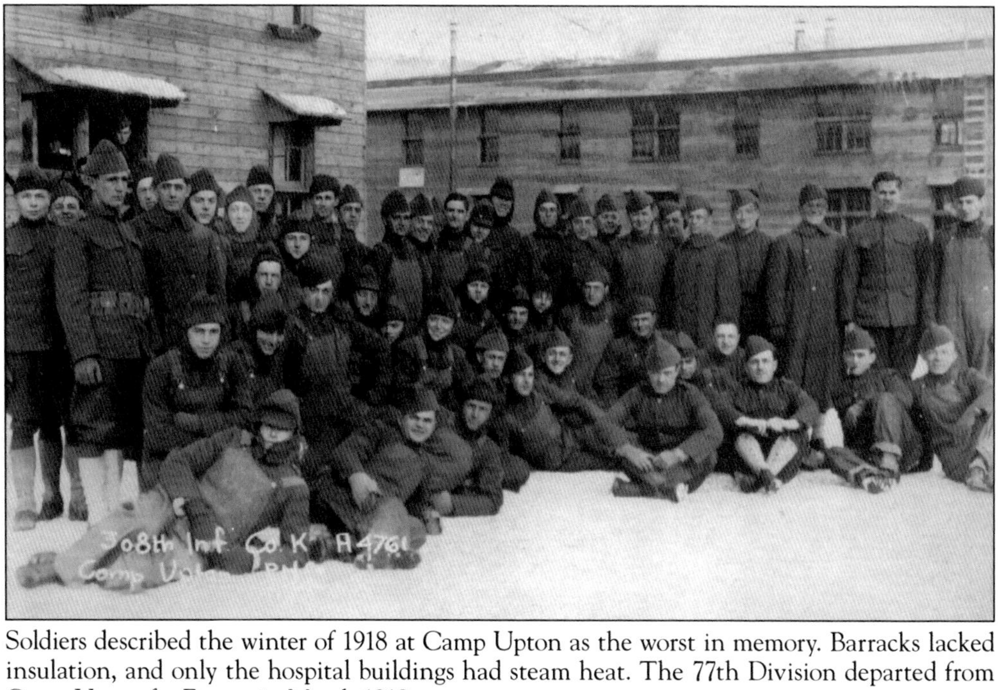

Soldiers described the winter of 1918 at Camp Upton as the worst in memory. Barracks lacked insulation, and only the hospital buildings had steam heat. The 77th Division departed from Camp Upton for France in March 1918.

Four

YIP! YIP! YAPHANK

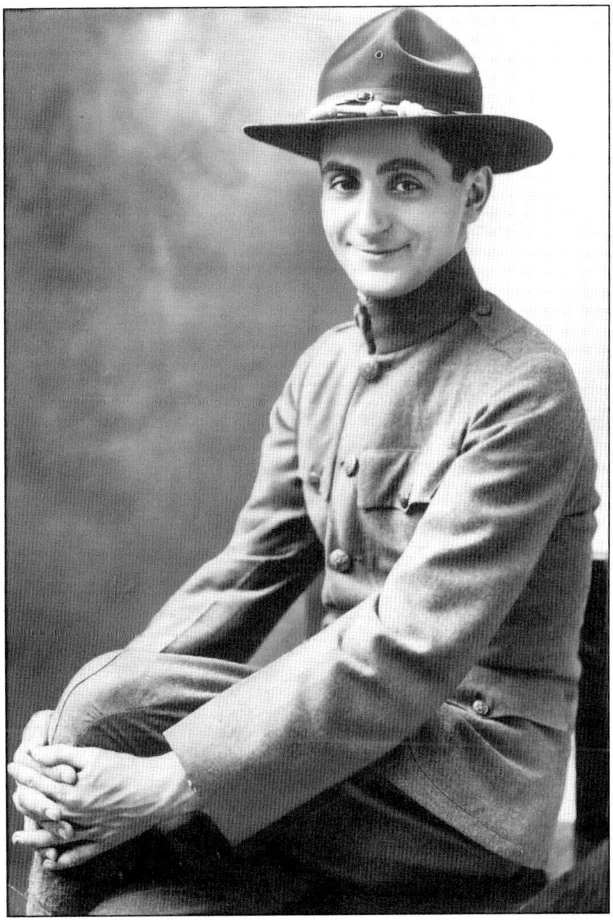

Noted Tin Pan Alley songwriter Irving Berlin was drafted into the National Army in the spring of 1918 and sent to Camp Upton. Tin Pan Alley was located in New York City on Twenty-Eighth Street between Fifth and Sixth Avenues. Tunesmiths cranked out patriotic songs that had the whole country singing. Irving Berlin was the most well-known songwriter at the time. (Courtesy of the Library of Congress.)

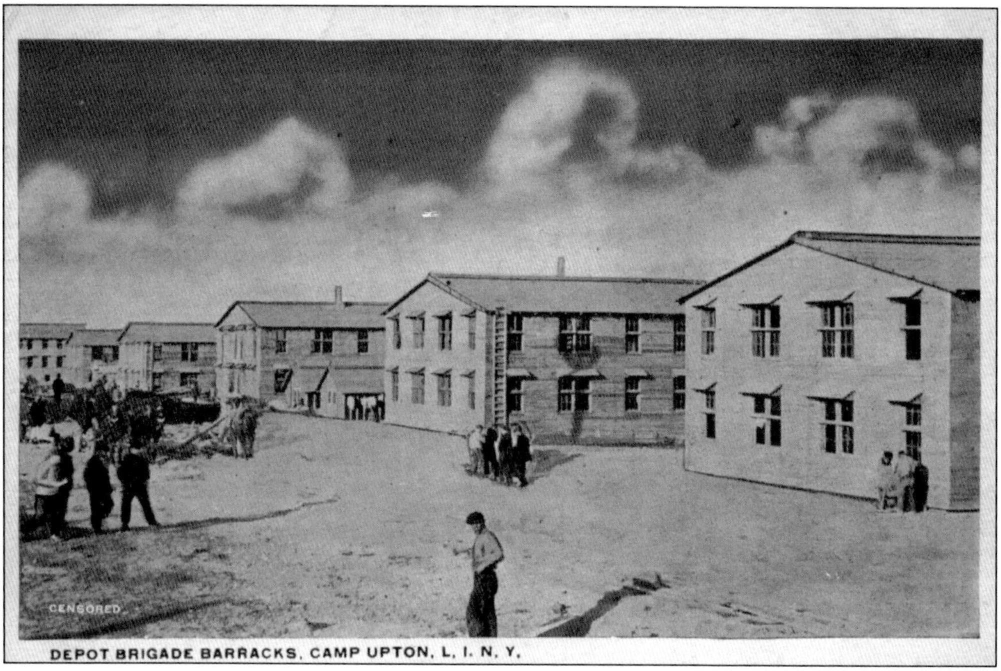

Irving Berlin, born in Russia as Israel Baline, became a US citizen in February 1918 and was surprised when he was summoned by the Army weeks later. New York newspapers proclaimed "US Army Takes Berlin!" When he first arrived at the camp, two local women visited General Bell and asked Berlin to change his name!

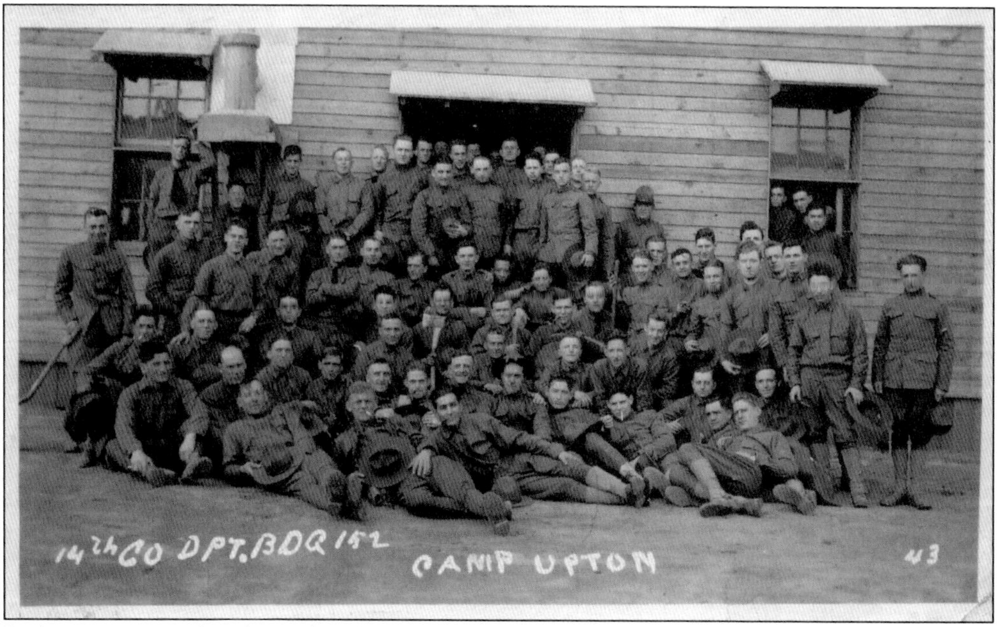

Berlin was assigned to the 20th Infantry, 152nd Depot Brigade, a training and receiving element of the 77th Division organized by General Bell. General Bell soon realized Berlin would be a better entertainer in camp than a soldier.

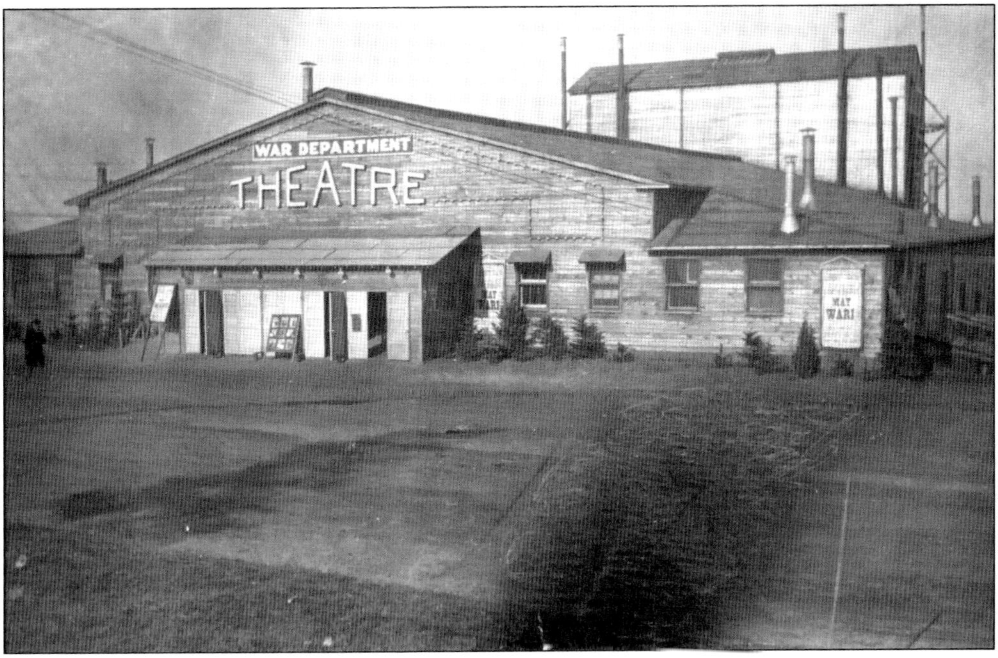

The role of depot brigades was to receive and organize recruits, provide them with uniforms, equipment, and initial military training, and then send them to France to fight on the front lines. Berlin convinced his superiors that rather than march and train, his role ought to be to write music, put on a show, and raise the morale of his fellow soldiers and inspire the country.

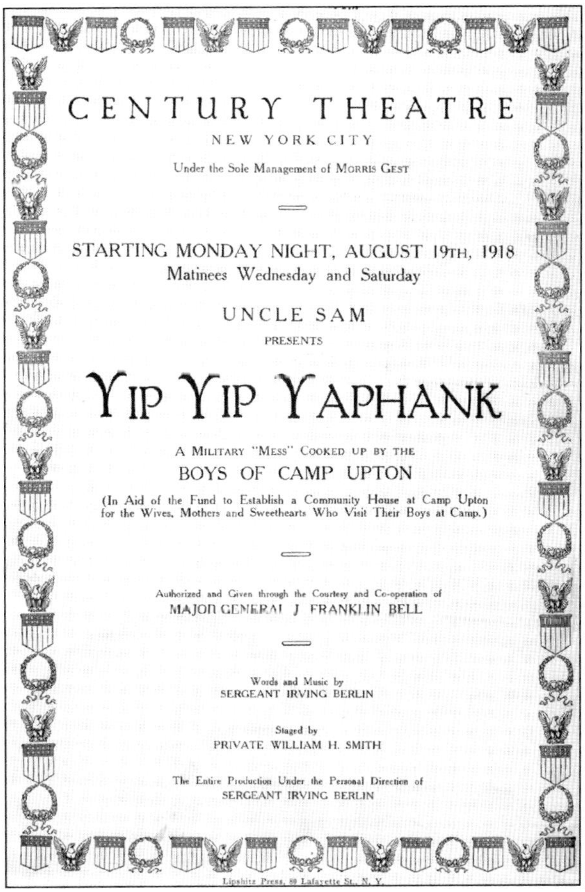

CENTURY THEATRE

NEW YORK CITY

Under the Sole Management of MORRIS GEST

STARTING MONDAY NIGHT, AUGUST 19TH, 1918

Matinees Wednesday and Saturday

UNCLE SAM

PRESENTS

YIP YIP YAPHANK

A MILITARY "MESS" COOKED UP BY THE

BOYS OF CAMP UPTON

(In Aid of the Fund to Establish a Community House at Camp Upton for the Wives, Mothers and Sweethearts Who Visit Their Boys at Camp.)

Authorized and Given through the Courtesy and Co-operation of

MAJOR GENERAL J. FRANKLIN BELL

Words and Music by

SERGEANT IRVING BERLIN

Staged by

PRIVATE WILLIAM H. SMITH

The Entire Production Under the Personal Direction of

SERGEANT IRVING BERLIN

Lipshitz Press, 80 Lafayette St., N. Y.

Berlin, accustomed to late hours as a songwriter in New York City, did not adjust well to reveille at Camp Upton. Legend has it General Bell elevated his rank to sergeant in exchange for writing and producing a show that would raise funds for a community house at Camp Upton. General Bell felt the camp needed a community house for public meetings and patriotic song rallies that would seat 5,000 to 10,000. Sergeant Berlin agreed to write a show, *Yip! Yip! Yaphank*, and he no longer had to report at reveille.

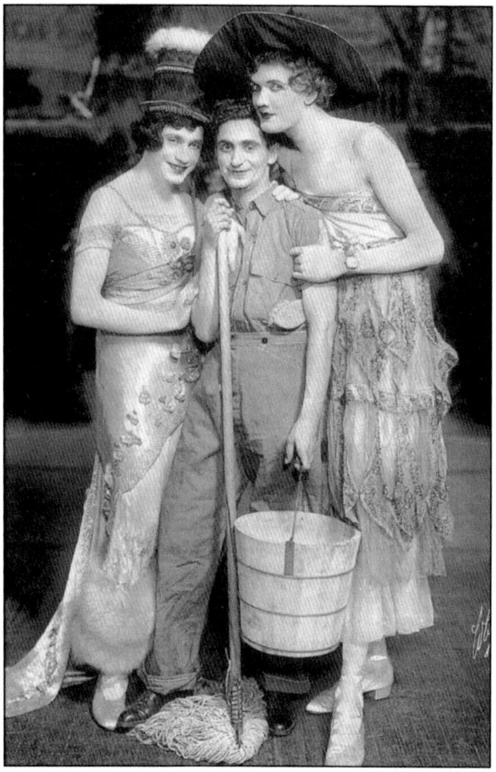

The entire production of *Yip! Yip! Yaphank*, with an all-soldier cast, including these two "ladies" with Berlin (he chose 350 soldiers from Camp Upton to perform), was under the personal direction of Sgt. Irving Berlin, had 32 performances, and made $83,000 for the community house General Bell wanted. However, the war ended, and the community house was never built.

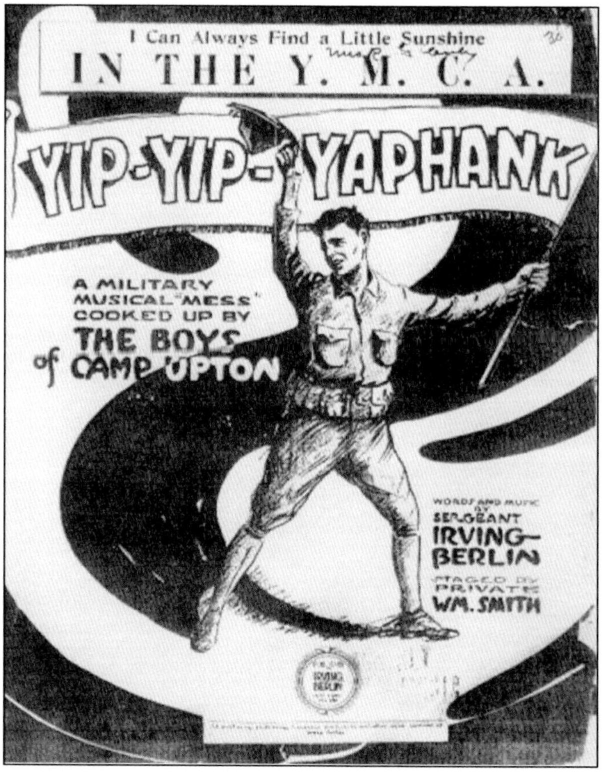

Songs in the show included "What a Difference a Uniform Will Make," "I Can Always Find a Little Sunshine in the YMCA," "Oh How I Hate to Get Up in the Morning," and "We're on Our Way to France." General Bell attended the opening night performance and stated, "I have heard that Berlin is among the foremost songwriters of the world, and now I believe it. . . . Berlin is as good a soldier as he is a songwriter and as popular in Camp Upton as he is on Broadway."

The cover of the program for *Yip! Yip! Yaphank* featured an illustration by Archie Gunn. Noted for his sentimental and humorous drawings, he was born in England and died in Garden City on Long Island.

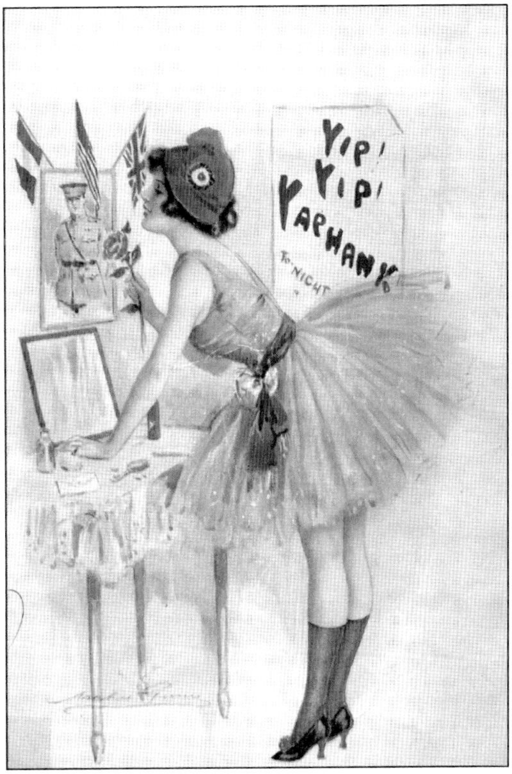

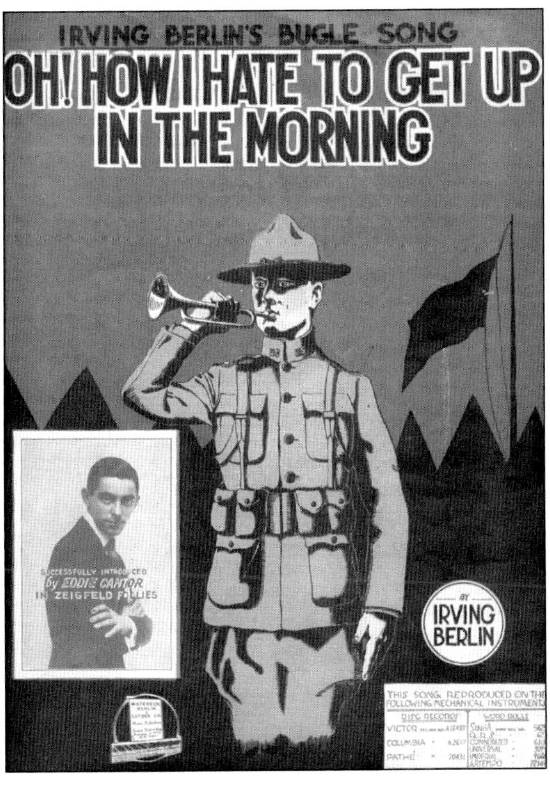

The sheet music of "Oh How I Hate to Get Up in the Morning" included the dedication "to my friend Private Howard Friend who occupies the cot next to mine and feels as I do about the bugler." Taps was at 10:00 p.m., much too early for night owl Irving Berlin.

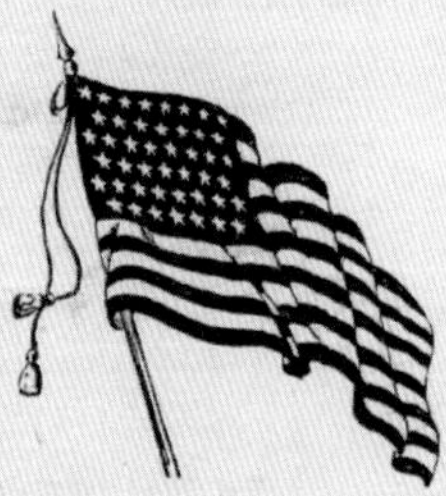

CENTURY TH

Under the Sole Management of - - - - - - - - - -

Starting Monday Evening,

"YIP YIP YA

A Musical "Mess" Cooked up by The Bo

Words and Music by Sergeant Irving

Staged by Private Will H. Smit

CHORUS

Privates:

ABBOTT
BREEN
BECKER
BRENNICK
BJORNQUIST
BRESLIN
BALLING
BARNETT
BARSALION
B. BROWN
BRAGG
BOLTON
D. BROWN
A. BROWN
BOYLE
BOLLES
BUCKLEY
BURNS
BORGRAFF
BURNETT
BREYER
BRYDE
CAMPBELL
COLANDO
CULLEN
CONDEN
CONWAY

Privates:

CLINE
CRONIN
HAGGERTY
M. CLARK
CHURCHILL
CAHILL
CONDNEN
DERMODY
DEGNAN
DEICK
DINAN
DAVISON
DONALDSON
ELLIS
EPSTEIN
EWELL
FRANCOIS
FREDERICH
FEILBERMAN
FITZPATRICK
FROST
GEDNEY
GOLD
GAULEN
GROSSMAN
GARBLICK
GINSBERG

ACT ONE:

Scene 1 — UPTON MILITARY MINS

In Command - - - - - - - - - Captain Pa

Interlocutor - - - - Q. M. Sergeant Major W

End Men - - - - Privates Hughie L. Clark

Harry Green

CHORUS

Privates:

GIGLIO
GORMAN
GOLDOFF
GRAY
GOLENSKI
GILLEN
DOWNEY
GILMAN
HAYES
HERMAN
HEIM
HILL
HOLLY
JORGAS
JACOBS

Privates:

JOHNSON
JOHNSTON
JONAS
KOUCH
KRUGER
KADER
KARSCH
JACK KELLY
SGT. JOE KELLY
J. KELLY
First Sgt. E. KELLY
KIME
KUBLER
LONG
LEDERER

Tambourine Drill - - - - - - - - - - - - - - - - Entire Company

Opening Chorus, "Hello, Hello, Hello" - - - - - - - - - - Entire Company

Solos - - - - - - - - - - - - - - Prvts. Brennan, Brenna, Phillips, Snyder, Schor and Ward

The Captains Address to the Company: - - - - - - - - - - - - - - -

Program Continued

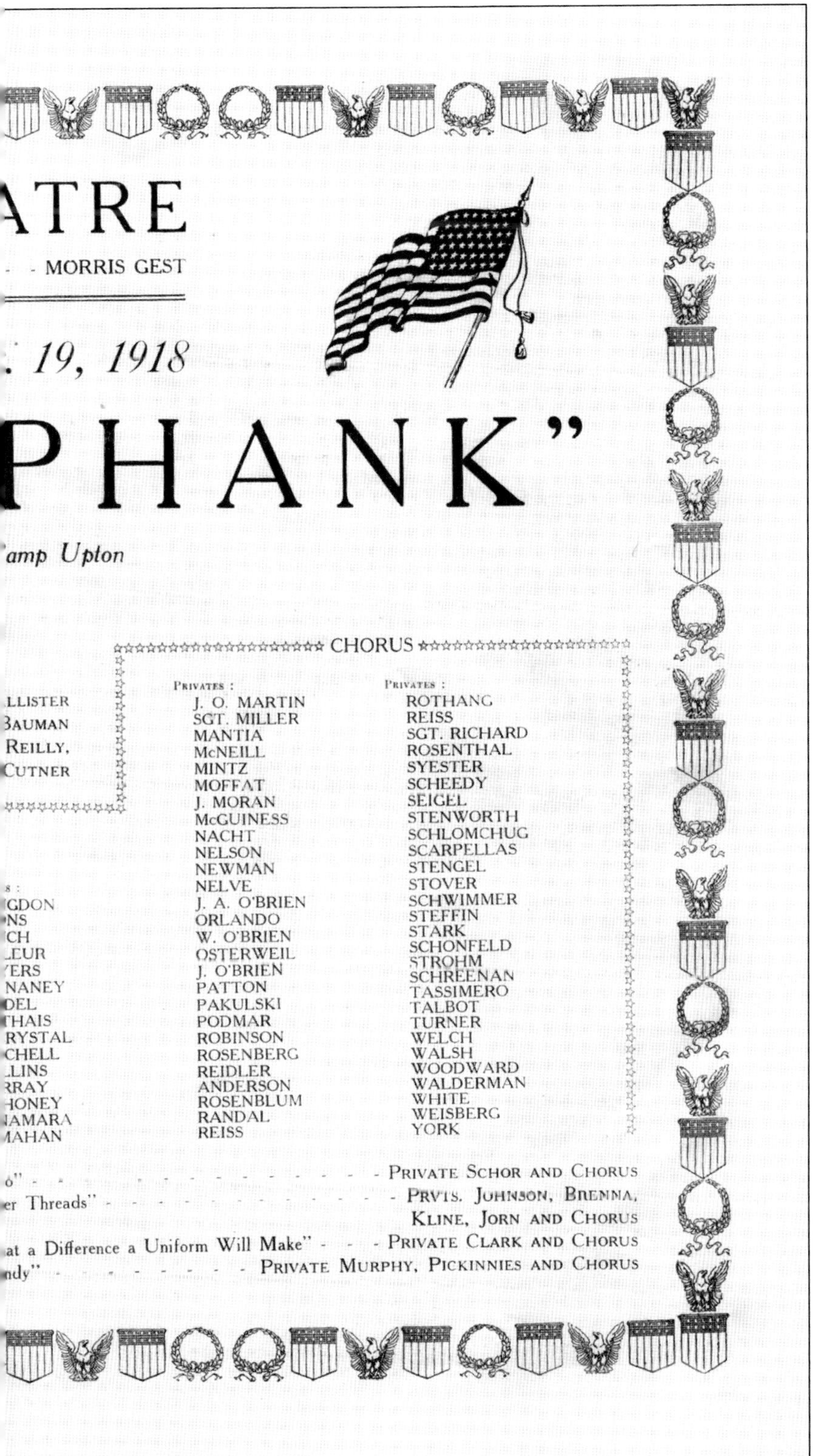

ATRE

- - MORRIS GEST

. 19, 1918

PHANK"

'amp Upton

CHORUS

	Privates:	Privates:
LLISTER	J. O. MARTIN	ROTHANG
BAUMAN	SGT. MILLER	REISS
REILLY,	MANTIA	SGT. RICHARD
CUTNER	McNEILL	ROSENTHAL
	MINTZ	SYESTER
	MOFFAT	SCHEEDY
	J. MORAN	SEIGEL
	McGUINESS	STENWORTH
	NACHT	SCHLOMCHUG
	NELSON	SCARPELLAS
	NEWMAN	STENGEL
s:	NELVE	STOVER
GDON	J. A. O'BRIEN	SCHWIMMER
NS	ORLANDO	STEFFIN
CH	W. O'BRIEN	STARK
LEUR	OSTERWEIL	SCHONFELD
ERS	J. O'BRIEN	STROHM
NANEY	PATTON	SCHREENAN
DEL	PAKULSKI	TASSIMERO
THAIS	PODMAR	TALBOT
RYSTAL	ROBINSON	TURNER
CHELL	ROSENBERG	WELCH
LINS	REIDLER	WALSH
RRAY	ANDERSON	WOODWARD
HONEY	ROSENBLUM	WALDERMAN
AMARA	RANDAL	WHITE
AHAN	REISS	WEISBERG
		YORK

o" - - - - - - - - - - - - - Private Schor and Chorus
er Threads" - - - - - - - - - - - - - Prvts. Johnson, Brenna, Kline, Jorn and Chorus
at a Difference a Uniform Will Make" - - - Private Clark and Chorus
ndy" - - - - - - - - Private Murphy, Pickinnies and Chorus

Yip! Yip! Yaphank, a military musical "mess" cooked up by the boys of Camp Upton in aid of the fund to establish a community house at Camp Upton for the wives, mothers, and sweethearts who visit their boys at camp, opened on August 19, 1918, and played for one month at the Century Theatre on Broadway. After one month, the show moved to the larger Lexington Theatre from September 2 to 14, 1918. The Century Theatre, built in 1909 by the noted firm of Carrere & Hastings, faced Central Park and was demolished in 1926.

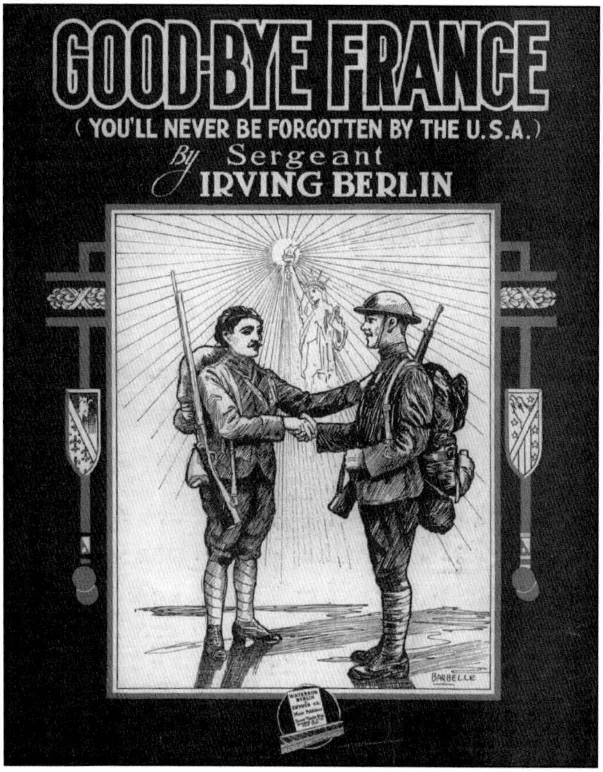

Later patriotic songs penned by Berlin included "Goodbye France" and "We Won the War to End the Wars." In 1938, Berlin's songs from *Yip! Yip! Yaphank* were featured in the movie *Alexander's Ragtime Band*, starring Alice Faye, Tyrone Power, and Don Ameche. (Courtesy of Fred Sietz.)

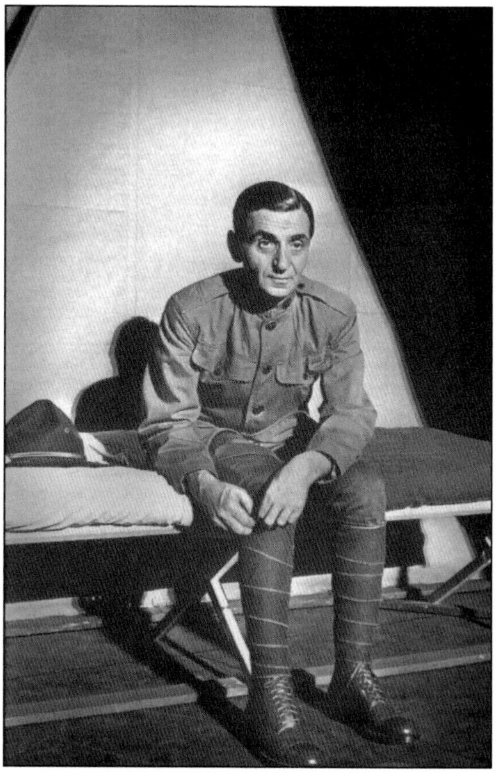

In 1941, Berlin created a new revue titled *This Is the Army*, which was basically a revival of *Yip! Yip! Yaphank*. Berlin selected 300 soldiers for the show at the reopened Camp Upton, where they rehearsed in April, May, and June. The show was a benefit for the Army Emergency Relief Fund, which helped needy soldiers and their families in the event of an illness or a death in the family. The show opened on July 4, 1941, at the Broadway Theatre and was ultimately taken around the world to entertain American troops, earning $10 million in its three-year run. Berlin called it "artistically the greatest experience of his life."

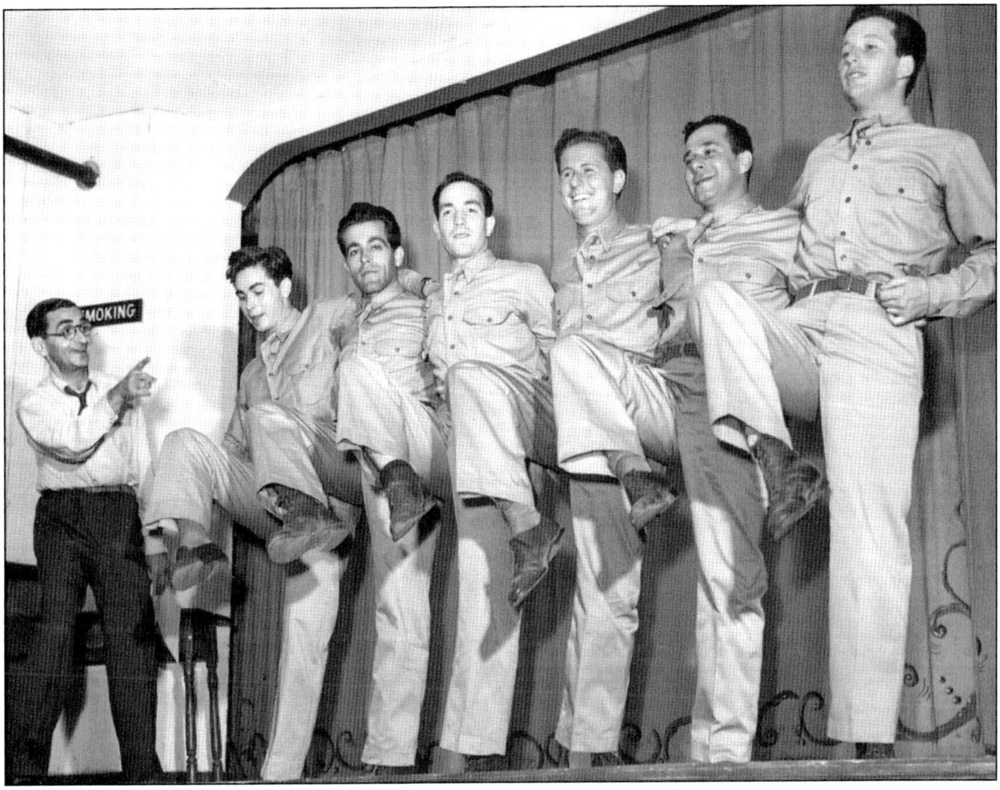

These World War II soldiers who will act as the "girlies" rehearse a dance act in *This Is the Army*. After rehearsing at Camp Upton, the company took the show around the world for three years. Berlin was asked to produce "a show that could be played on a field of combat." From left to right are Irving Berlin, Nelson Barclift, Bob Sidney, Fred Kelly, Fred Kapner, Larry Weill, and Allan Manson.

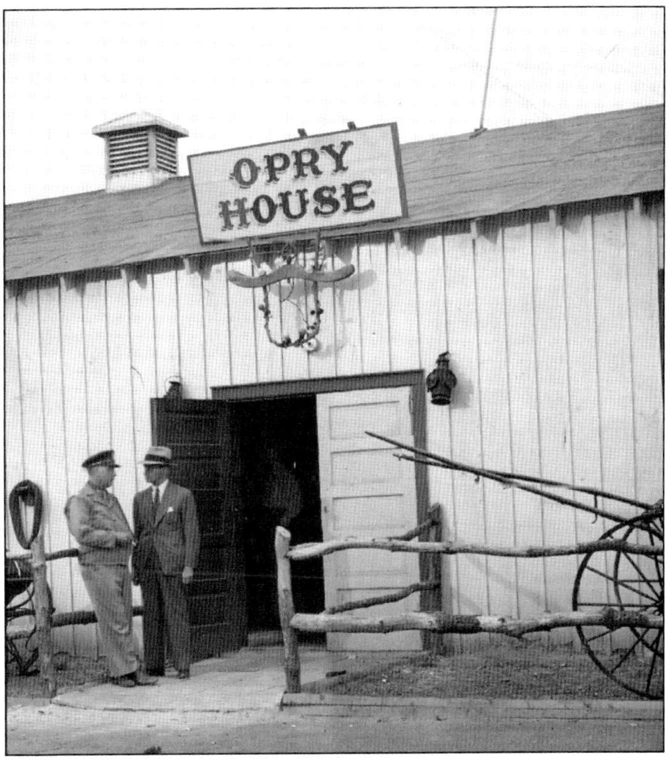

Berlin is shown here at Camp Upton in 1942 with Capt. A.H. Rankin, morale officer. The "Opry House" was formerly a 10-car garage.

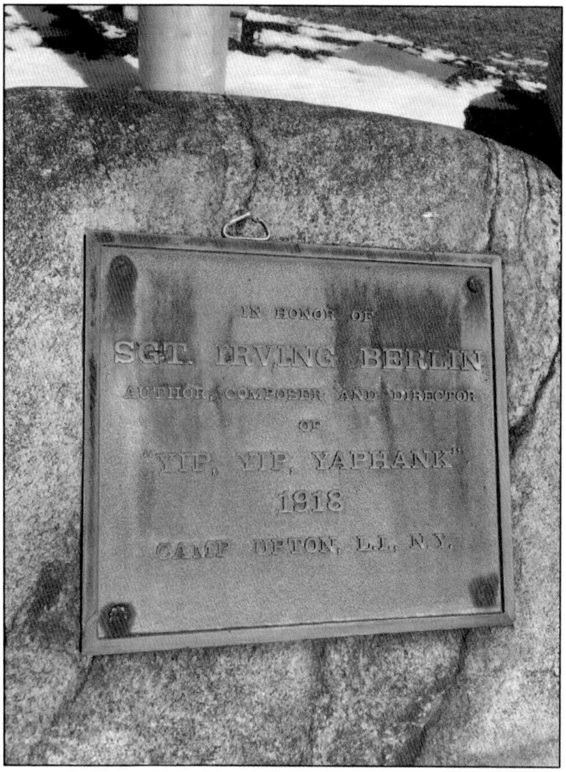

A plaque honoring Irving Berlin was placed at the former camp, now Brookhaven National Laboratory. In 1982, Berlin wrote to local town councilman Donald Zimmer: "I can't tell you how much this means to me. Camp Upton was a very important part of my life and having this plaque placed on the grounds of Brookhaven touches me deeply." Irving Berlin died on February 26, 1989, at the age of 101 and is buried in Woodlawn Cemetery.

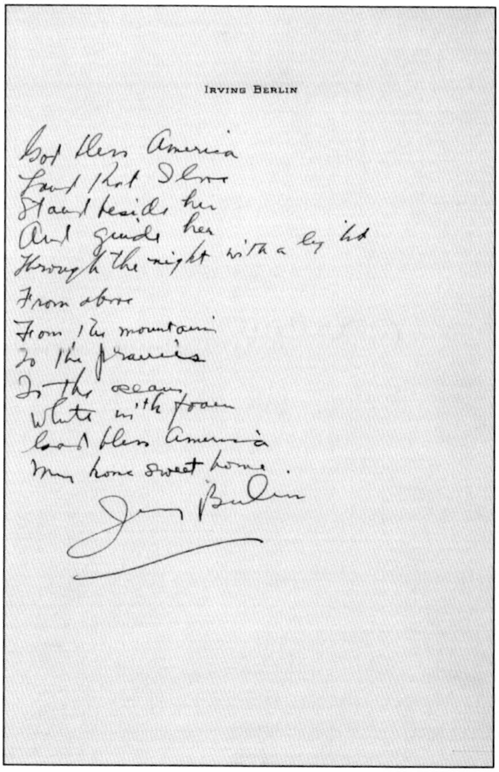

IRVING BERLIN

God bless America
Land that I love
Stand beside her
And guide her
Through the night with a light
From above
From the mountains
To the prairies
To the oceans
White with foam
God bless America
My home sweet home.
Irving Berlin

The song "God Bless America" had been written for *Yip! Yip! Yaphank,* but Berlin thought it too serious for the show and tucked it away. Twenty years later, Kate Smith asked him for a song for her radio program. Berlin remembered the song and changed a few lines, and Smith sang it for the first time on her show on the 20th anniversary of Armistice Day, November 11, 1938. Berlin signed all royalties for the song to the Boy Scouts and Girl Scouts of America. By 1950, it earned $200,000 for the God Bless America Fund, and it continues to generate millions in income for Scouting programs in impoverished areas. Berlin described it as the most important song he had ever written. (Courtesy of the Library of Congress.)

Five

Rebuild and Reuse

Armistice was declared on November 11, 1918, and the mission at Camp Upton changed. In 1919, the camp served as a demobilization center for returning veterans. By 1921, the entire camp was to be torn down and auctioned off. The entire contents of the camp were dismantled and sold at public auction in August 1921, yielding $360,000 for the War Department. A total of 1,660 buildings, utilities, and improvements were either dismantled or moved. Many dormitories and sheds were purchased by local farmers as outbuildings and cabins, with prices starting at $40. Buildings from Camp Upton were ported over the Great South Bay by a developer to the summer community of Cherry Grove on Fire Island. The hurricane of 1938 destroyed most of the little town, but some original homes remain. Other buildings were moved to the surrounding towns of Yaphank, Middle Island, Coram, and Rocky Point. The mess hall at Baiting Hollow Boy Scout Camp is believed to originate from Camp Upton.

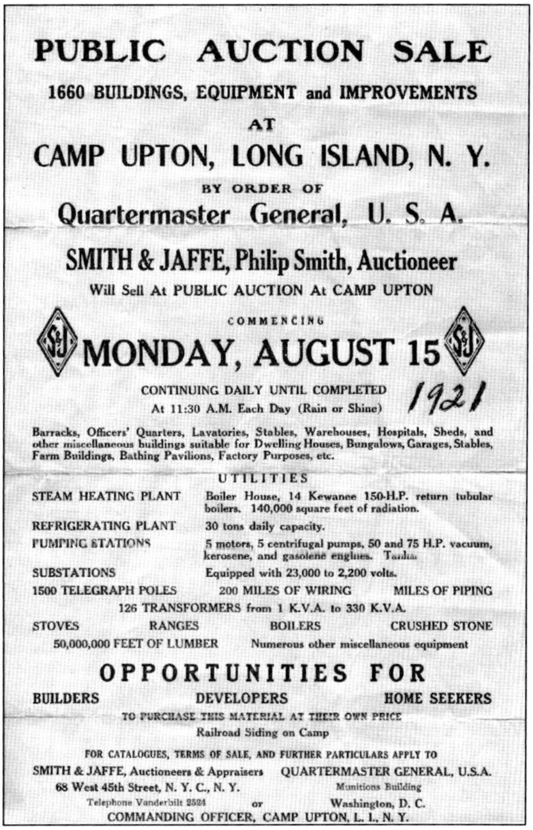
PUBLIC AUCTION SALE

1660 BUILDINGS, EQUIPMENT and IMPROVEMENTS

AT

CAMP UPTON, LONG ISLAND, N. Y.

BY ORDER OF

Quartermaster General, U. S. A.

SMITH & JAFFE, Philip Smith, Auctioneer

Will Sell At PUBLIC AUCTION At CAMP UPTON

COMMENCING

MONDAY, AUGUST 15

CONTINUING DAILY UNTIL COMPLETED 1921

At 11:30 A.M. Each Day (Rain or Shine)

Barracks, Officers' Quarters, Lavatories, Stables, Warehouses, Hospitals, Sheds, and other miscellaneous buildings suitable for Dwelling Houses, Bungalows, Garages, Stables, Farm Buildings, Bathing Pavilions, Factory Purposes, etc.

UTILITIES

STEAM HEATING PLANT — Boiler House, 14 Kewanee 150-H.P. return tubular boilers. 140,000 square feet of radiation.

REFRIGERATING PLANT — 30 tons daily capacity.

PUMPING STATIONS — 5 motors, 5 centrifugal pumps, 50 and 75 H.P. vacuum, kerosene, and gasolene engines. Tanks

SUBSTATIONS — Equipped with 23,000 to 2,200 volts.

1500 TELEGRAPH POLES — 200 MILES OF WIRING — MILES OF PIPING

126 TRANSFORMERS from 1 K.V.A. to 330 K.V.A.

STOVES — RANGES — BOILERS — CRUSHED STONE

50,000,000 FEET OF LUMBER — Numerous other miscellaneous equipment

OPPORTUNITIES FOR

BUILDERS — DEVELOPERS — HOME SEEKERS

TO PURCHASE THIS MATERIAL AT THEIR OWN PRICE

Railroad Siding on Camp

FOR CATALOGUES, TERMS OF SALE, AND FURTHER PARTICULARS APPLY TO

SMITH & JAFFE, Auctioneers & Appraisers, 68 West 45th Street, N. Y. C., N. Y., Telephone Vanderbilt 2524

QUARTERMASTER GENERAL, U.S.A., Munitions Building, Washington, D. C.

or

COMMANDING OFFICER, CAMP UPTON, L. I., N. Y.

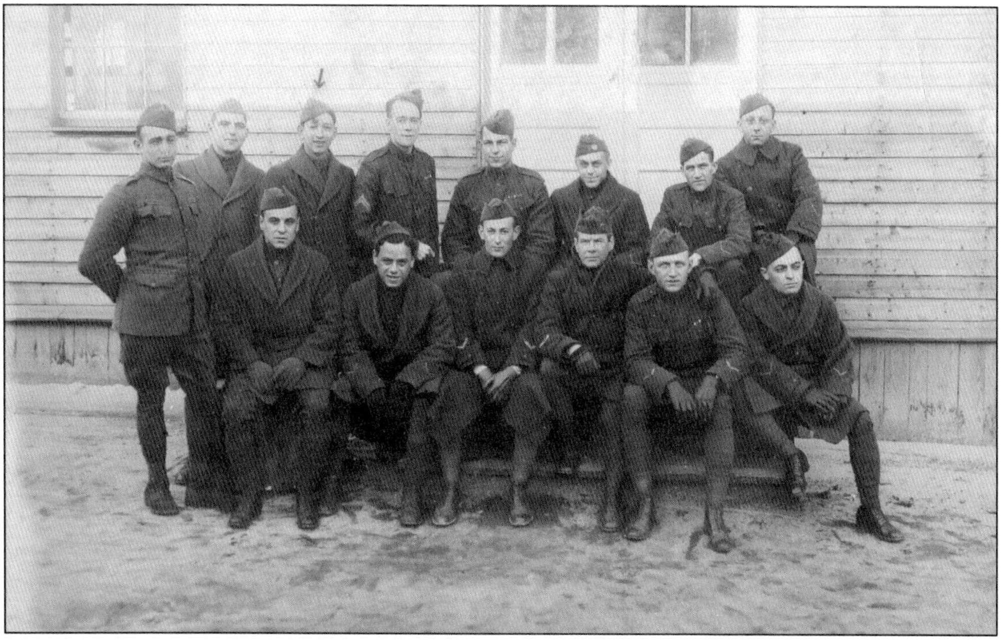

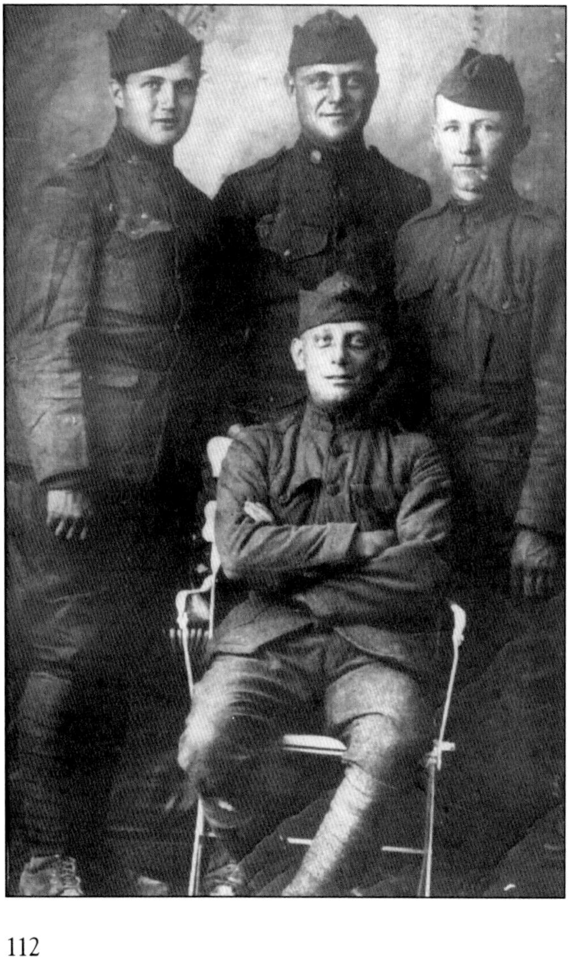

The uniforms these soldiers are wearing have wound and service chevrons indicating they have gone to the front, were wounded, and returned to recuperate at Camp Upton. They labeled this January 1919 photograph "The All A.W.O.L. Club." Several have service ribbons displayed on their jackets. Most signed the back of the photograph: Martin J. Schast, Louis, Jack, Harold Becker, James Howard, Albert J. Lowe, Tony DePrima, Leo Bigelow, S.J. Devine, and Arthur Rodriguez. Their close comradeship is apparent.

John E. Lynch, center, standing, was honorably discharged as a corporal in the 27th Division, 105th Infantry, in April 1919 after demobilizing at Camp Upton. He had enlisted in the first enlistment period on June 8, 1917, and served in France and Belgium including the engagement at Vierstratt Ridge, East Poperinghe Line, Hindenburg line, and St. Souplet. His cousin Jack Walsh stands to his left. (Courtesy of William S. Hale.)

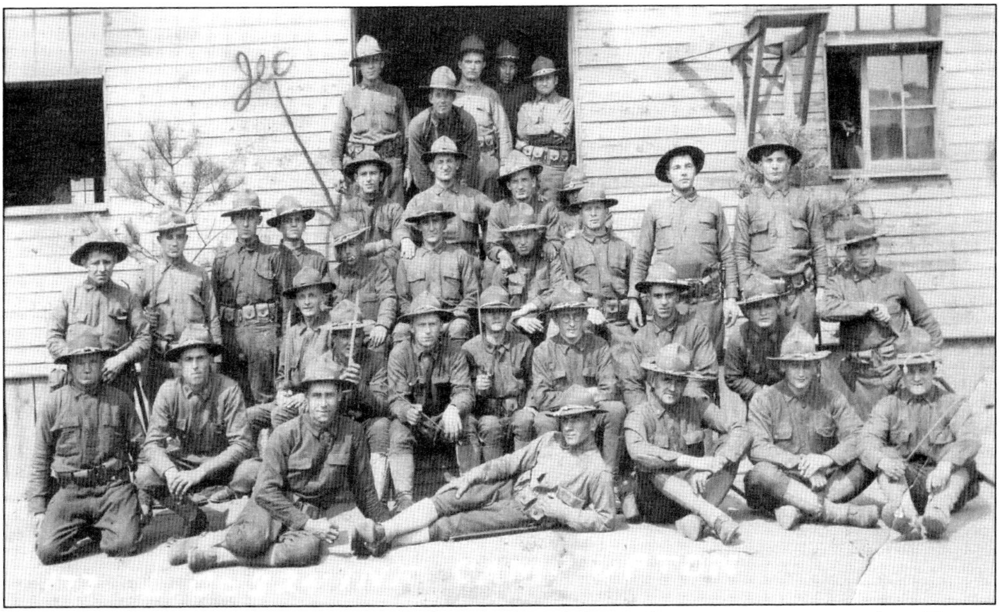

Camp Upton's location made it an ideal place to demobilize military units returning from France during the spring of 1919. These included the 27th Division, which was made up of New York National Guard Troops in early April, Upton's own 77th Division in early May, and the 81st "Lincoln" Division in late May, of which the 325th Infantry was a part.

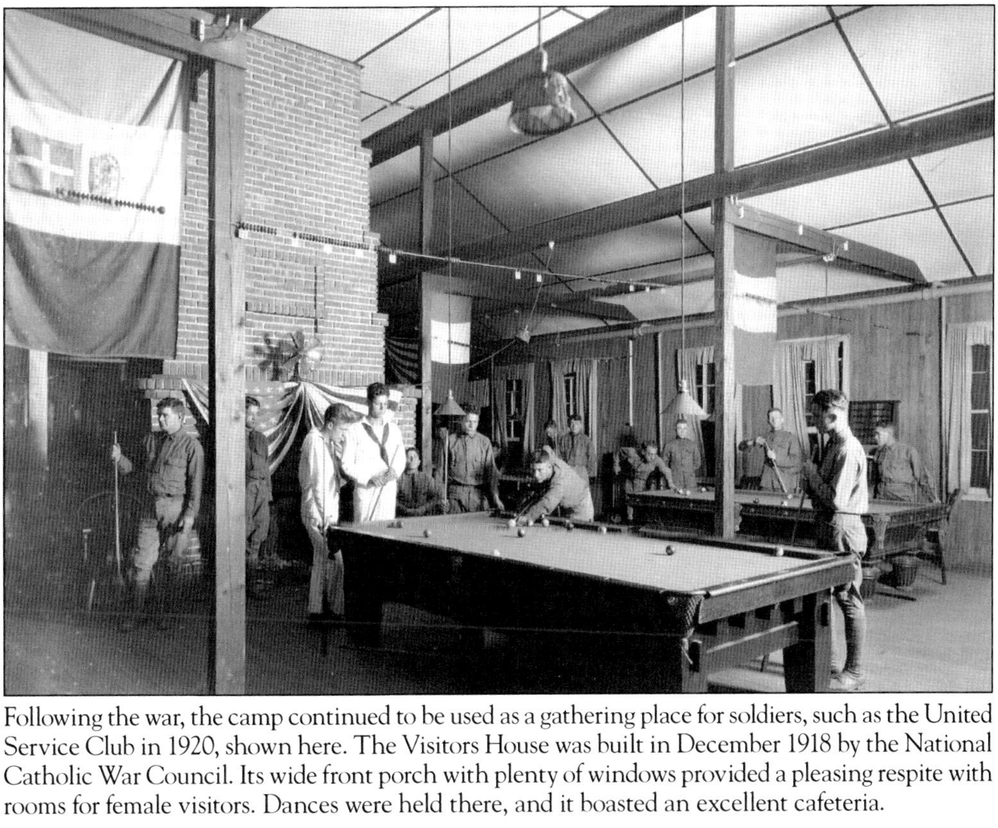

Following the war, the camp continued to be used as a gathering place for soldiers, such as the United Service Club in 1920, shown here. The Visitors House was built in December 1918 by the National Catholic War Council. Its wide front porch with plenty of windows provided a pleasing respite with rooms for female visitors. Dances were held there, and it boasted an excellent cafeteria.

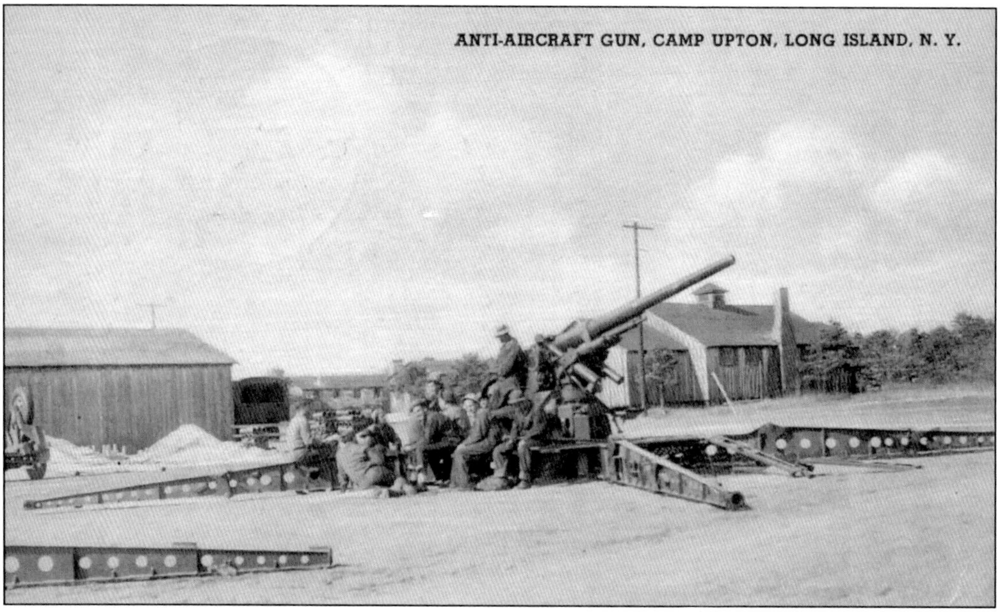

From 1923 to 1933, the 198th Coast Artillery used the vacant camp. It had organized in 1921 in Delaware and was redesignated from the Delaware National Guard in 1924 as an antiaircraft regiment. The 198th Coast Artillery was the first to arrive at the reopened Camp Upton in September 1940 prior to the US entry into World War II. Soldiers had to sleep in tents, as the buildings were not yet finished. Eventually, 1,200 Guardsmen would be housed there.

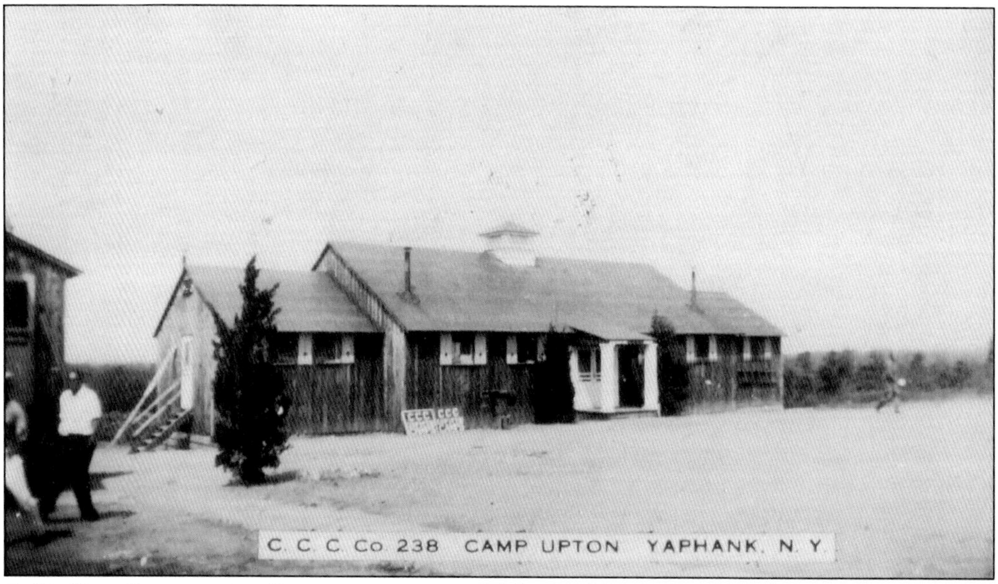

A Civilian Conservation Corps camp was established at Upton in 1933. The CCC was a federal program to put young men to work during the Great Depression. The workers were paid $1 a day. Four companies of 200 men built fire roads and planted two million pine and locust trees on the site. Acres were planted with seed-bearing grasses with the ultimate goal that the property would serve as a public game preserve. The men also assisted in fighting fires in the surrounding communities. Although the camp closed in 1938, hunting was not allowed in the fall of 1939. This is the only remaining CCC building.

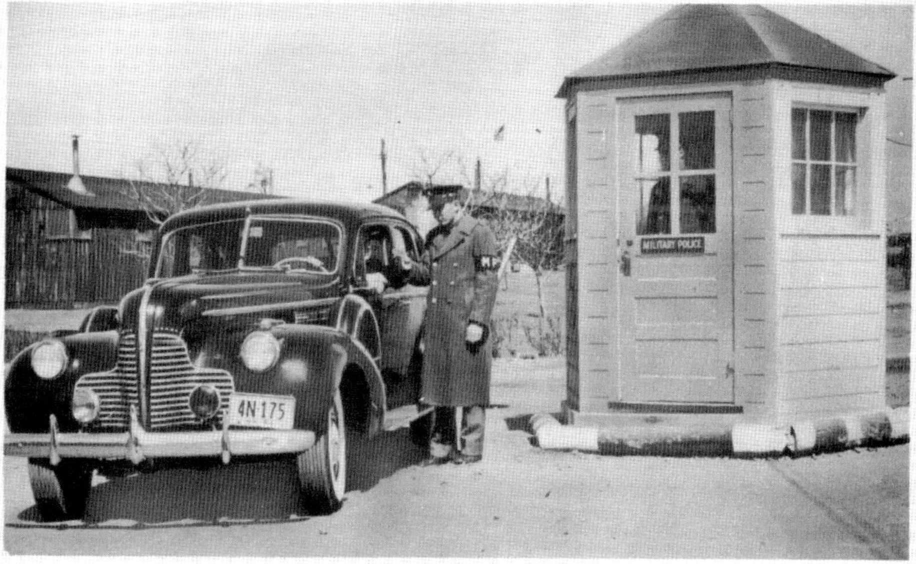

MILITARY POLICE DIRECTING TRAFFIC AT ENTRANCE, CAMP UPTON, L. I.

As a new war approached in 1940, it was necessary to utilize the former training camp once again to prepare troops for combat. The mile-and-a-half railroad spur from Yaphank was rebuilt after being removed 21 years earlier.

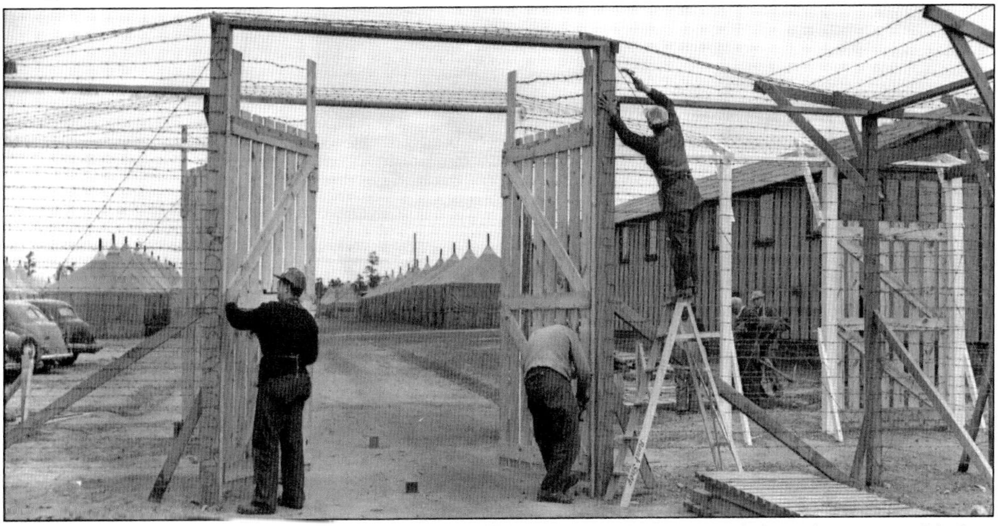

News photographic press releases in October 1941 described a new and disturbing use added to the camp. "Brand new alien internment nearing completion at Camp Upton, winterized tents to house 700 aliens," declared one story, then, "First camp of its kind to be set up in these parts since the world war." Within days of declaring war, the US government began taking custody of citizens of foreign countries. Seven hundred aliens would be housed in tents inside a 480,000-square-foot stockade surrounded by double walls of barbed wire for the duration of the war. Sentries would keep their charges under supervision night and day. In February 1942, a report by the Swiss legation stated that were 218 nationals of enemy countries in the camp, of whom 118 were Germans, 86 Japanese, and 14 Italians. Of these, 45 were from the crew of the German freighter *Odenwald*. However, their stay was short, and the detainees were shipped to Camp Meade in Maryland.

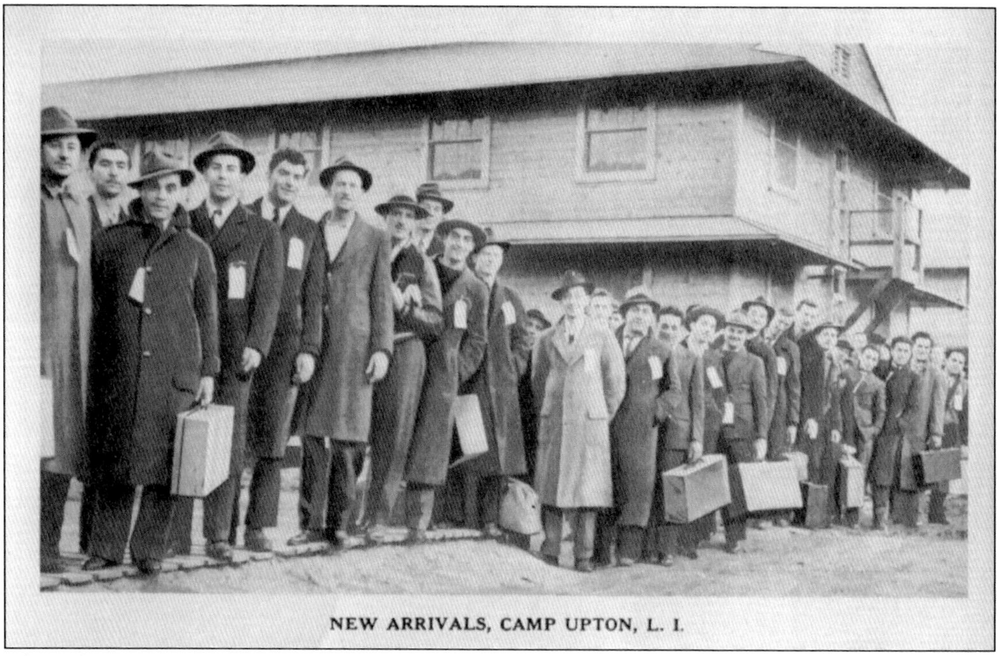

The Selective Training and Service Act, passed in September 1940, required all men ages 21 to 35 to register. The rebuilt camp served as an induction center for World War II soldiers, who generally stayed for six days before departing for Europe.

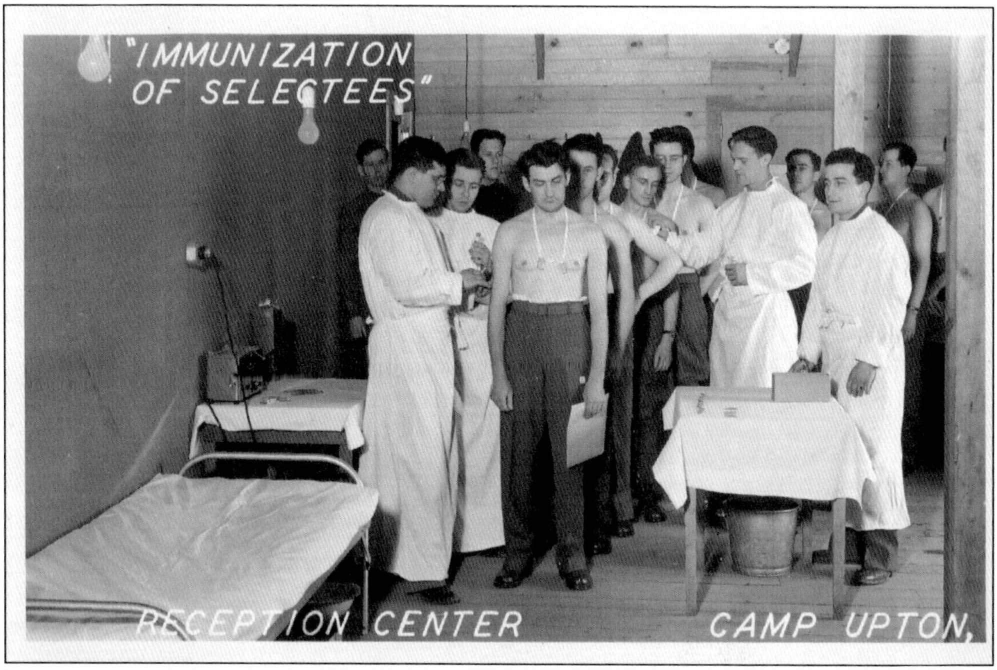

Immunizations were an important first stop for the recruits. Fighting in the Pacific would expose soldiers to malaria and typhoid. Vaccines were also provided to guard against measles, mumps, rubella, and tuberculosis. This war was the first major conflict in which fewer troops died of disease than battle injuries and wounds.

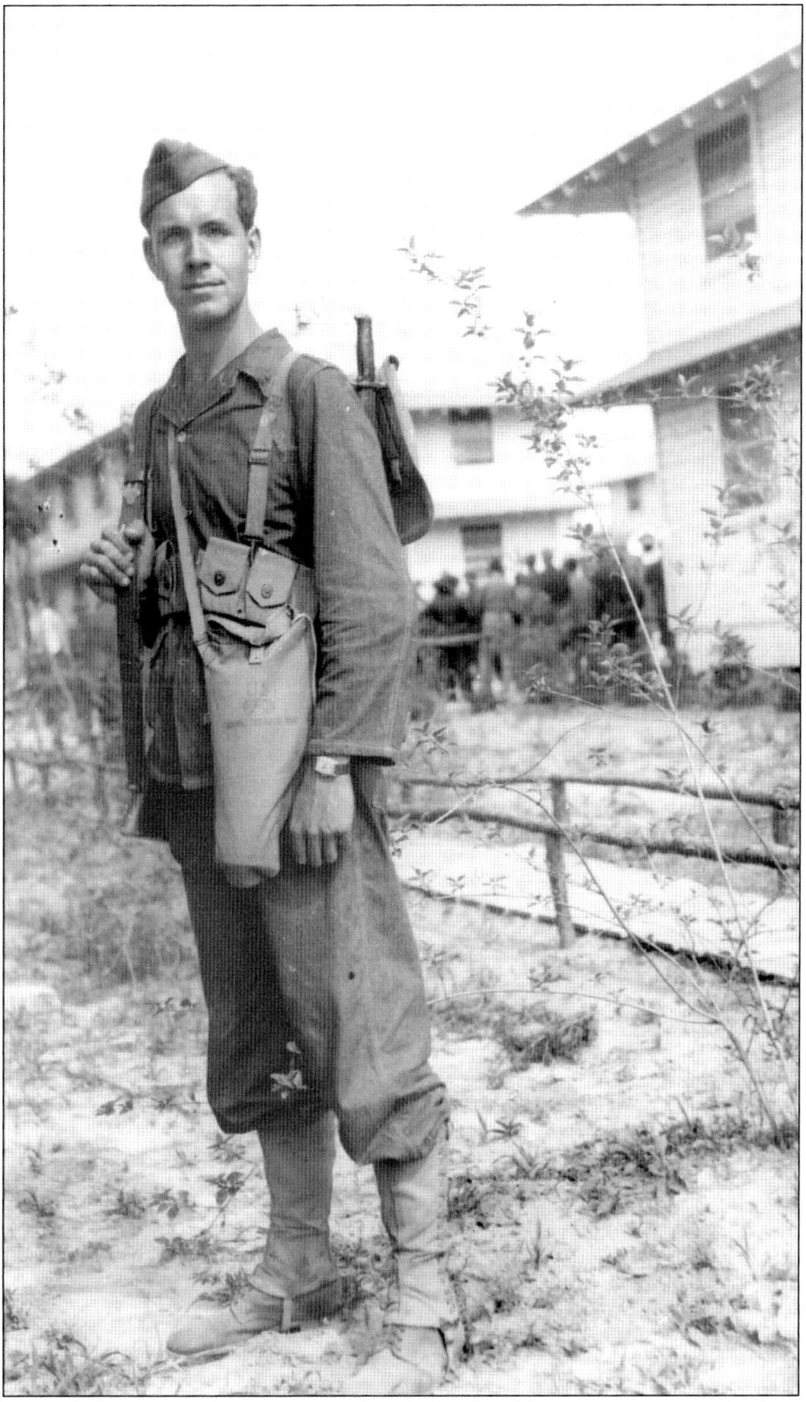

This photograph, labeled "Summer 1941 Bernard O'Rielly First Training BN," shows Private O'Rielly (perhaps a misspelling of O'Reilly) wearing a typical World War II HBT (herringbone twill cotton) fatigue jacket and pants, canvas leggings, issue boots, and an overseas cap. He is carrying a gas mask, haversack, webbing suspenders, cartridge belt, and rifle. (Courtesy of Brookhaven National Laboratory.)

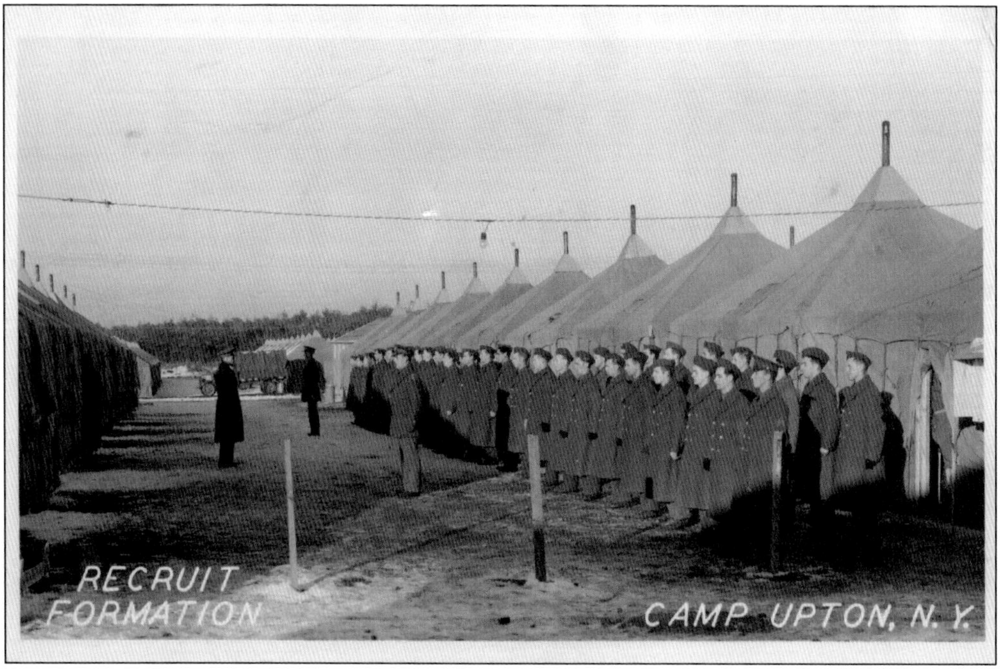

Recruits form in front of their tents. Col. Clair W. Baird was the commanding officer of the camp. More than 20,000 men and women came through from 1941 to 1945.

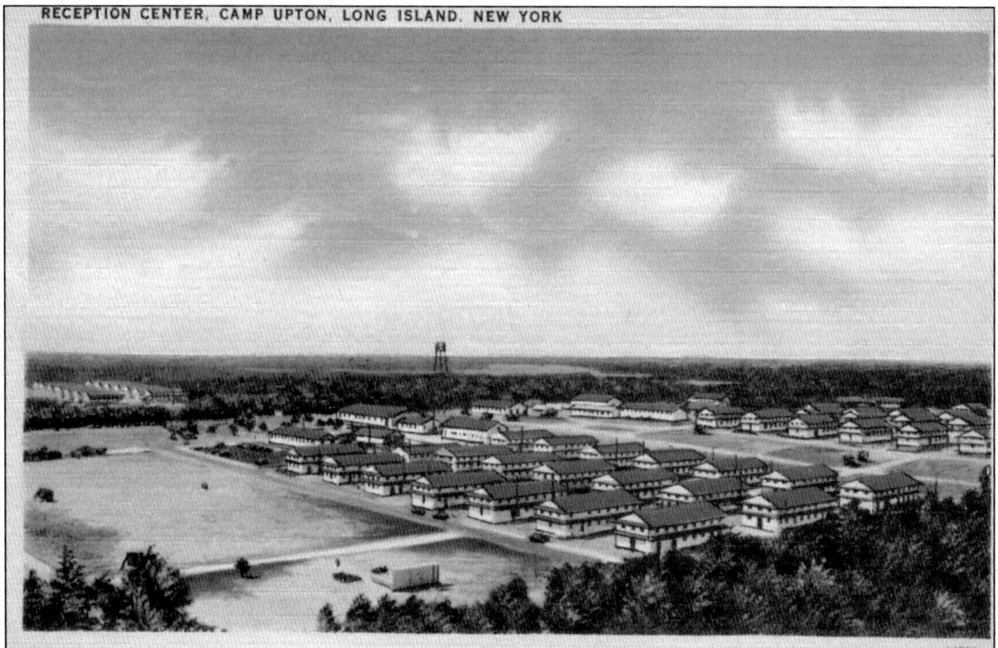

The reception center consisted of 56 buildings, including 40 barracks, 2 mess halls, an infirmary, a recreation hall, and 2 post exchanges.

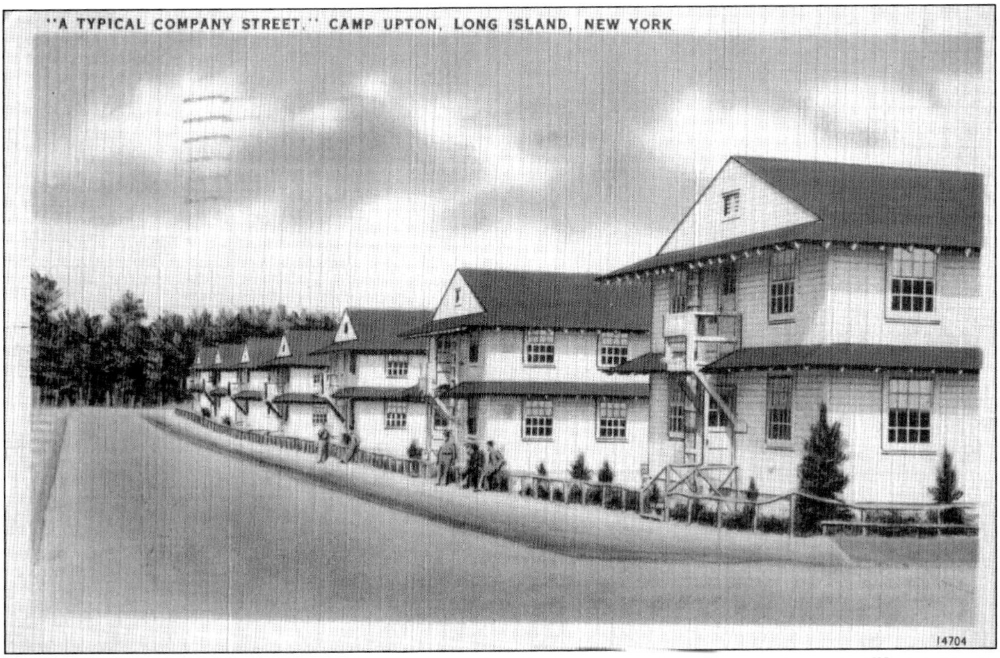

Two-story barracks were built similar to those that had been built and torn down in World War I, although they were smaller and could accommodate about 60 men. There were separate mess halls. Only a few of these buildings remain.

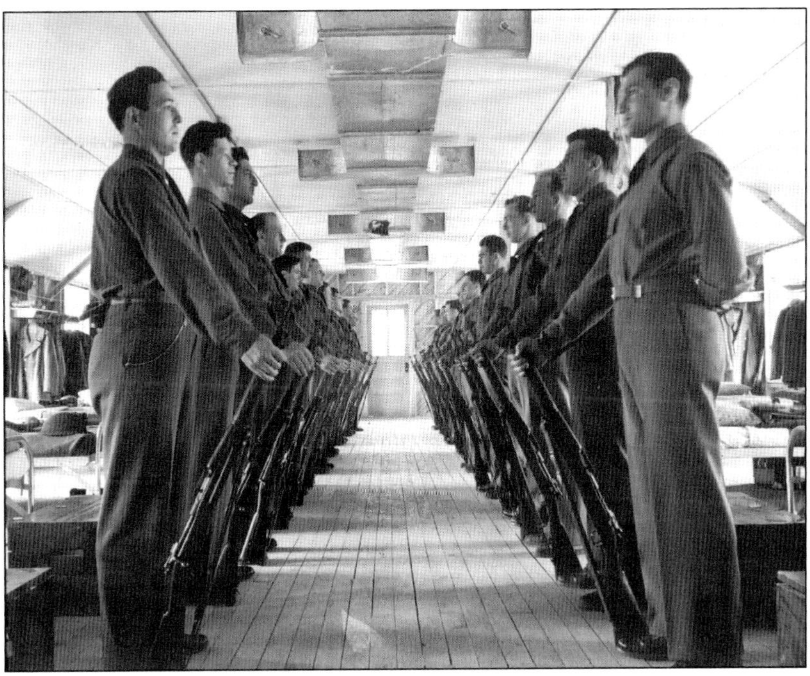

The interior of the World War II–era Quonset barracks can be seen in this photograph. One advantage over the 1917 buildings was central heating and plumbing.

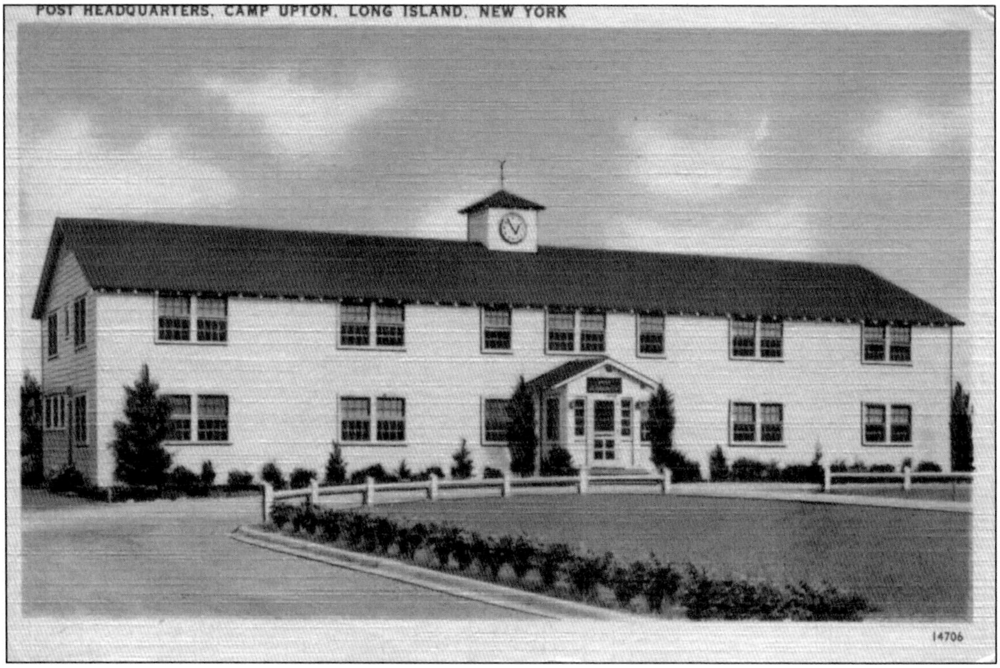

The former camp commander's headquarters building from World War II is now Building 50, Brookhaven National Laboratory Police Headquarters. The building and clock tower have remained unchanged. The reception center, built during the CCC years, is utilized as a banquet and conference center.

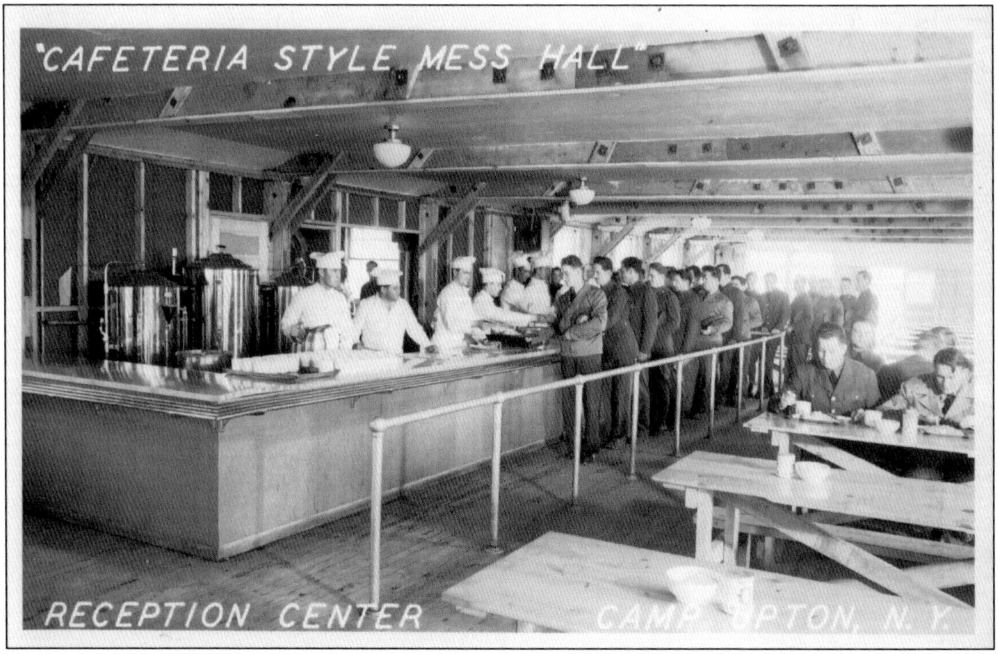

Meals were served in these cafeteria-style mess halls, as opposed to in the barracks buildings as they had been in World War I.

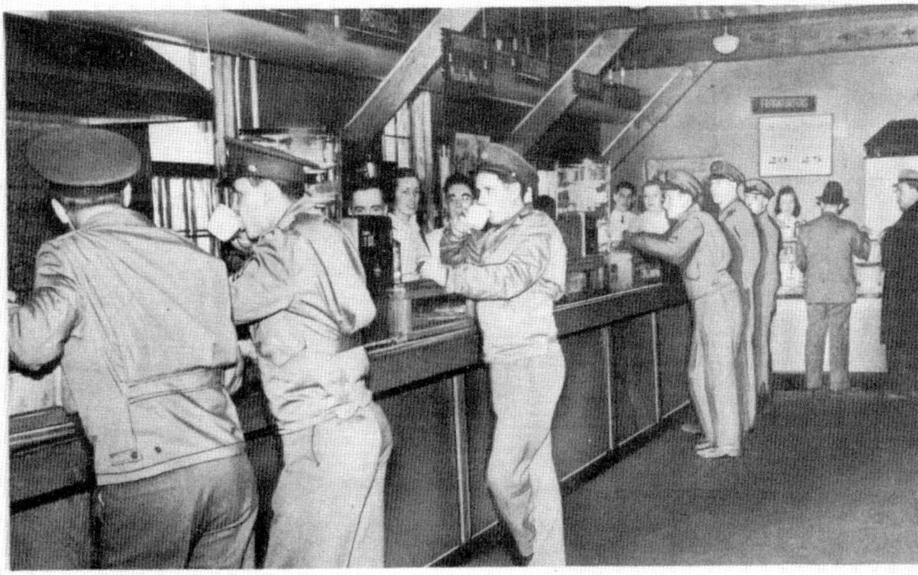

POST EXCHANGE RESTAURANT, CAMP UPTON, L. I.

A post exchange restaurant provided food and company for soldiers and visitors.

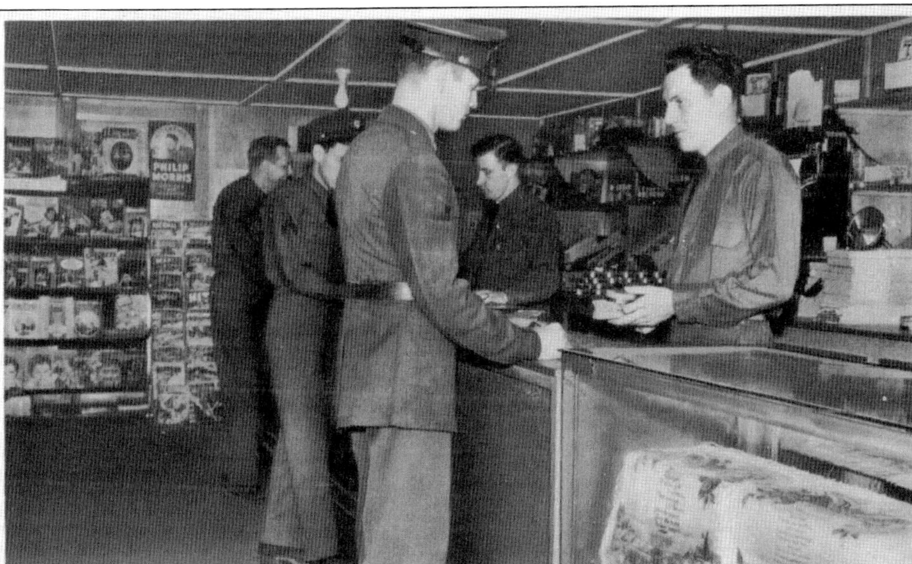

POST EXCHANGE, CAMP UPTON, L. I.

Soldiers could pick up the latest issue of *Upton Nooz*, Camp Upton's weekly newspaper, at the post exchange. The latest newspapers and magazines were also popular.

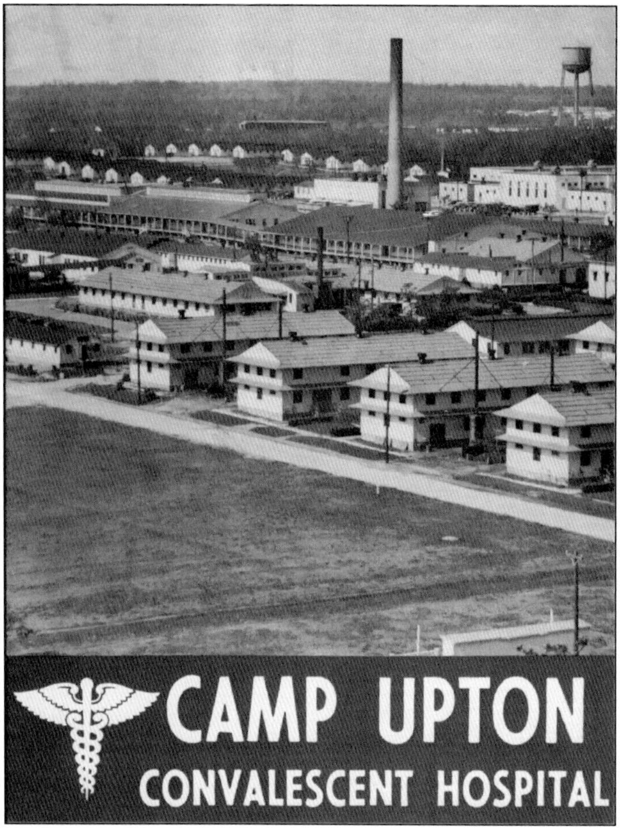

In September 1944, a convalescent and rehabilitation hospital was established to treat wounded war veterans. At its peak it had 3500 beds and treated more than 25,000 patients during its two years of operation. When it closed in June 1946, the equipment and supplies were sent to other Army posts and VA hospitals. The swimming pool and tennis courts remain from this era.

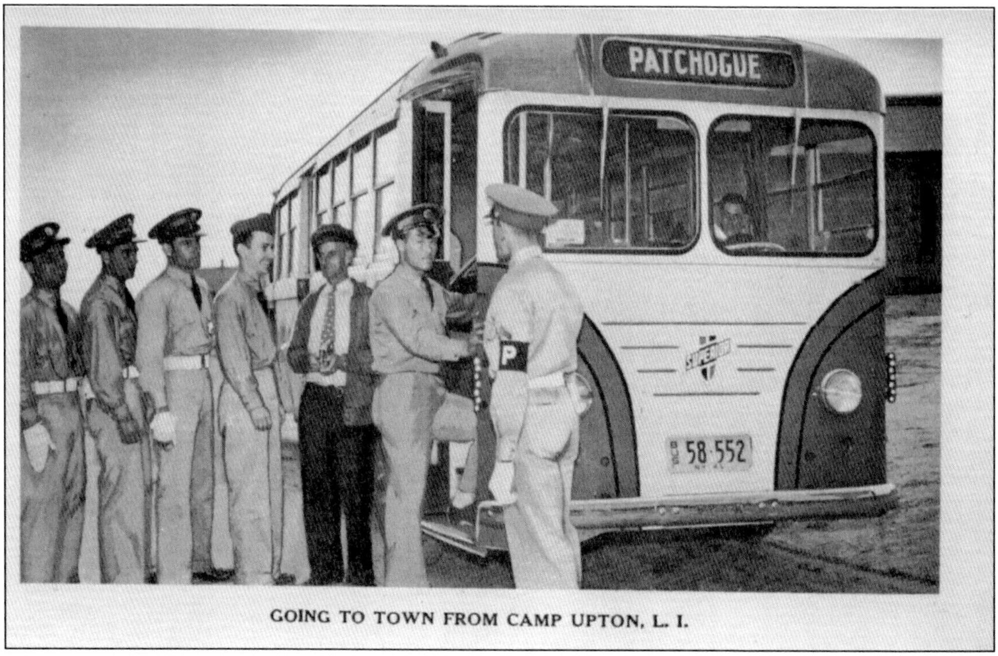

Recruits on furlough take the bus to Patchogue. The bus franchise was operated by Thomas J. Quinn.

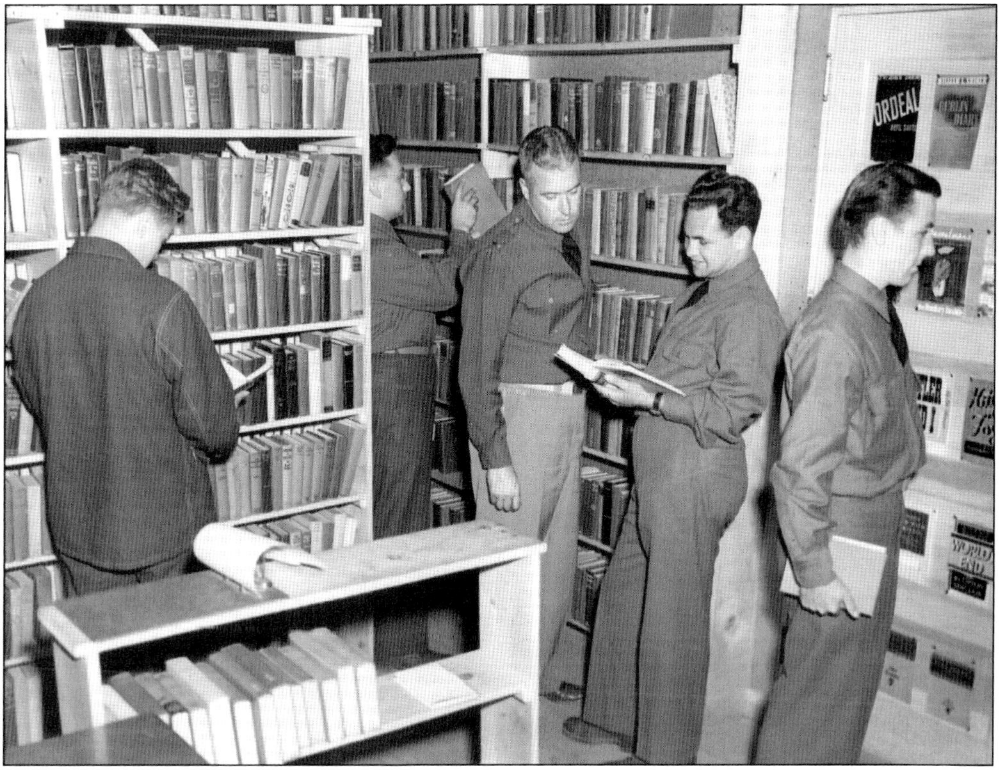

An Episcopal chaplain visits with recruits in the library at Camp Upton in March 1942.

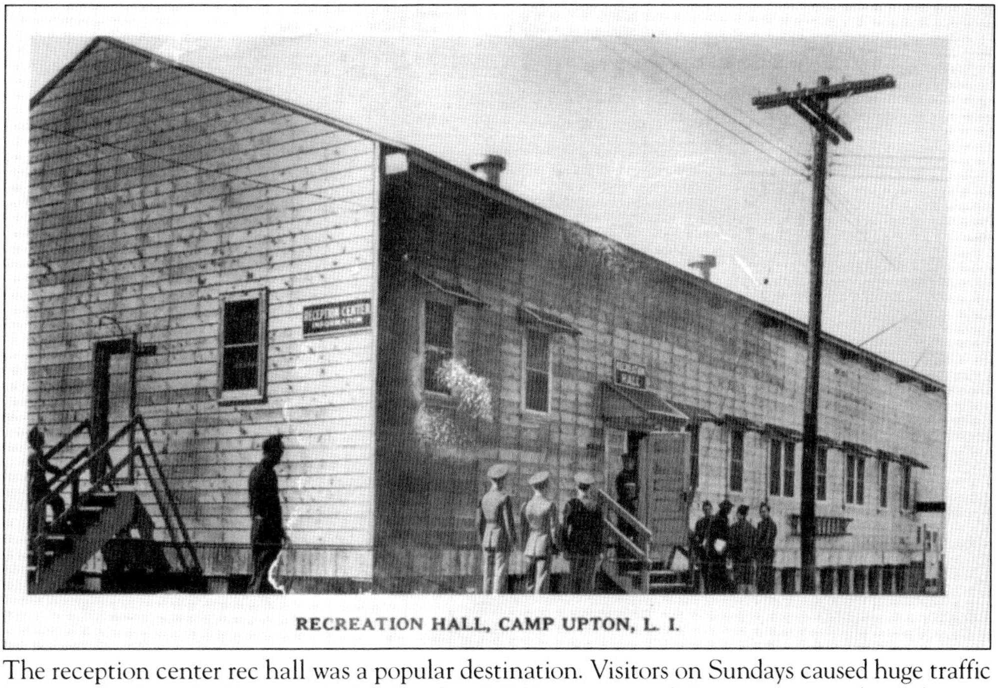

The reception center rec hall was a popular destination. Visitors on Sundays caused huge traffic jams on local roads. One particular Sunday, 3,500 cars entered the camp grounds, and many became mired in the parking fields.

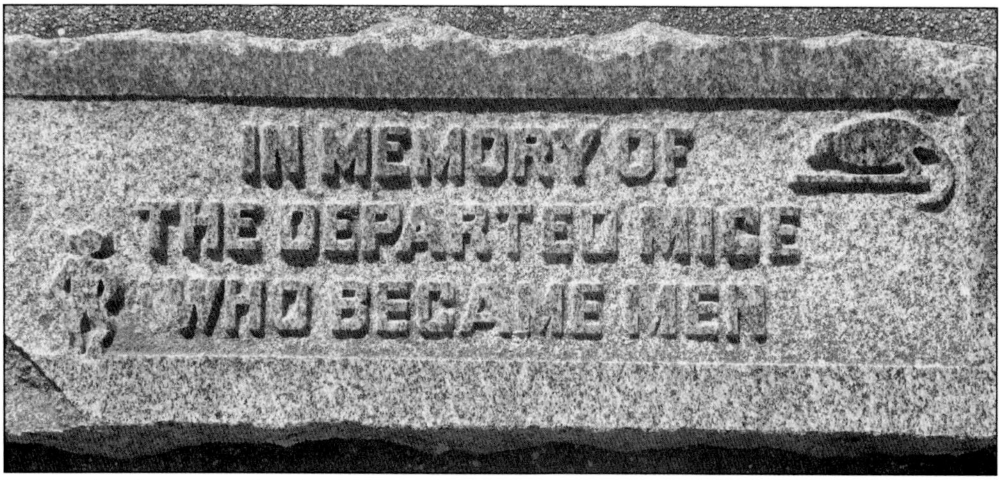

In 1944, local veterans placed a commemorative monument in front of the officers' club to honor their fellow soldiers and draftees who never made it overseas, referring to them as "mice." "Mice" was a nickname for the inductees when they first came to camp. In the 1950s, the marker was relocated to the police headquarters.

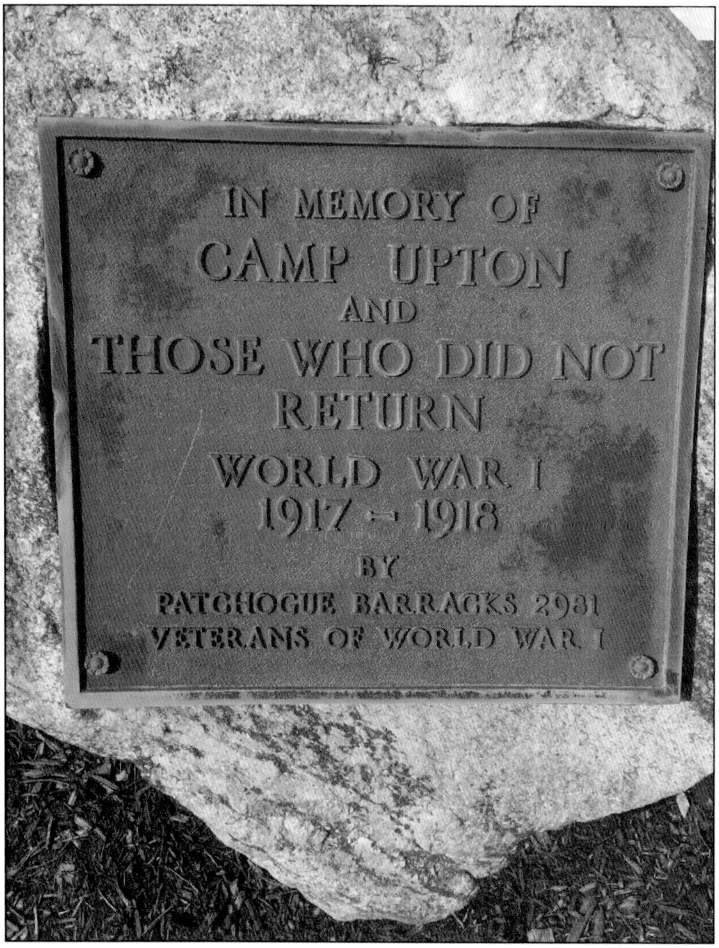

In 1976, local veterans from the Patchogue Barracks 2981 VFW Post rededicated the "mice" marker in its original location at the 1939 World's Fair flagpole with an additional marker recognizing the historical significance of Camp Upton in World War I.

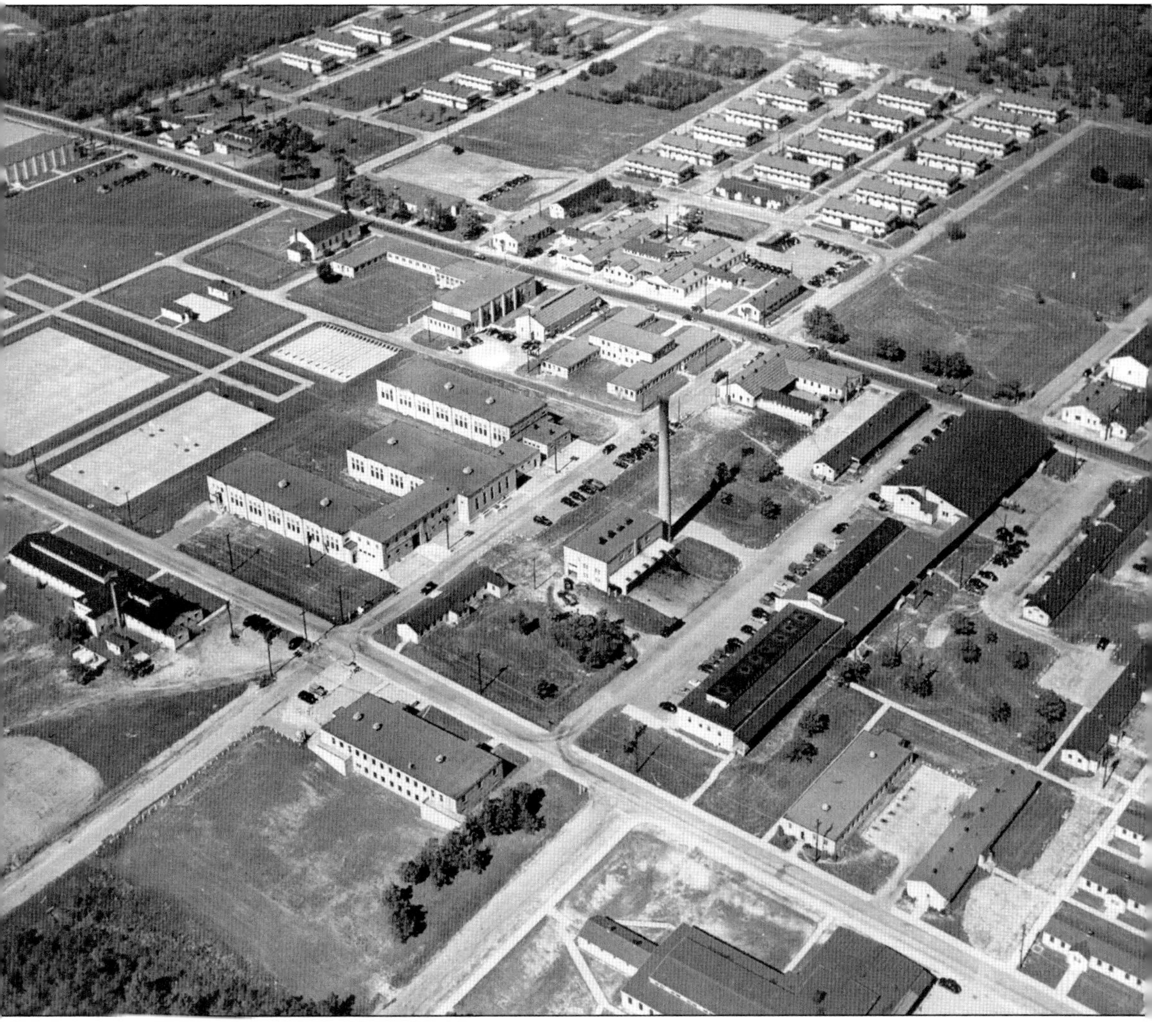

This aerial view of Camp Upton in 1948 shows the transition from military training camp to federal scientific laboratory. The powerhouse atop the original "Headquarters Hill" is clearly visible. On June 30, 1946, Camp Upton was officially declared surplus by the Army, which opened the way for a new era.

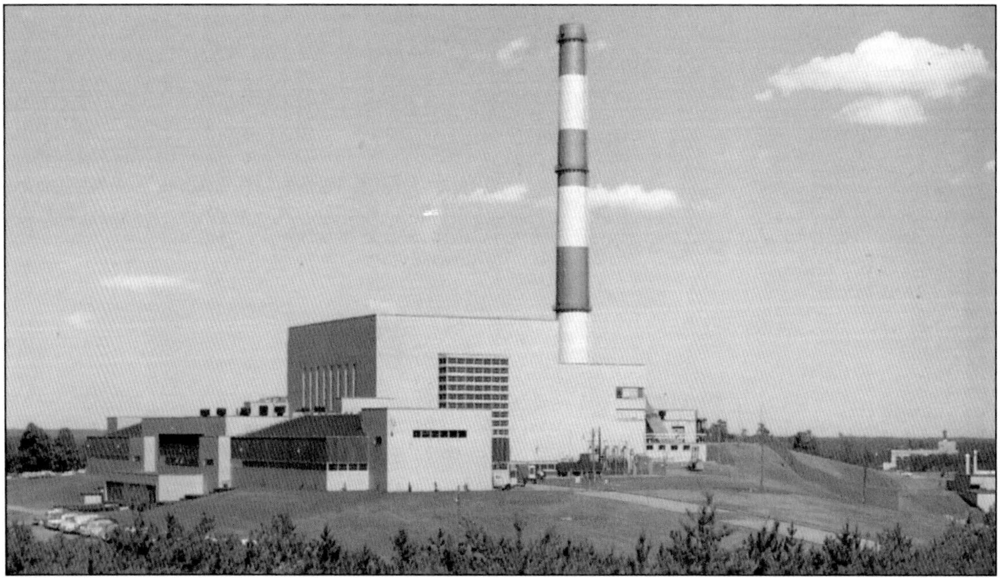

On March 1, 1947, the former site of Camp Upton officially became Brookhaven National Laboratory (BNL), the first national laboratory dedicated to discovering peaceful uses of the atom and the study of nuclear science. The site was chosen for its proximity to New York City and easy access to university scientists by car and rail. The former Army barracks from World War II were remodeled into laboratory space.

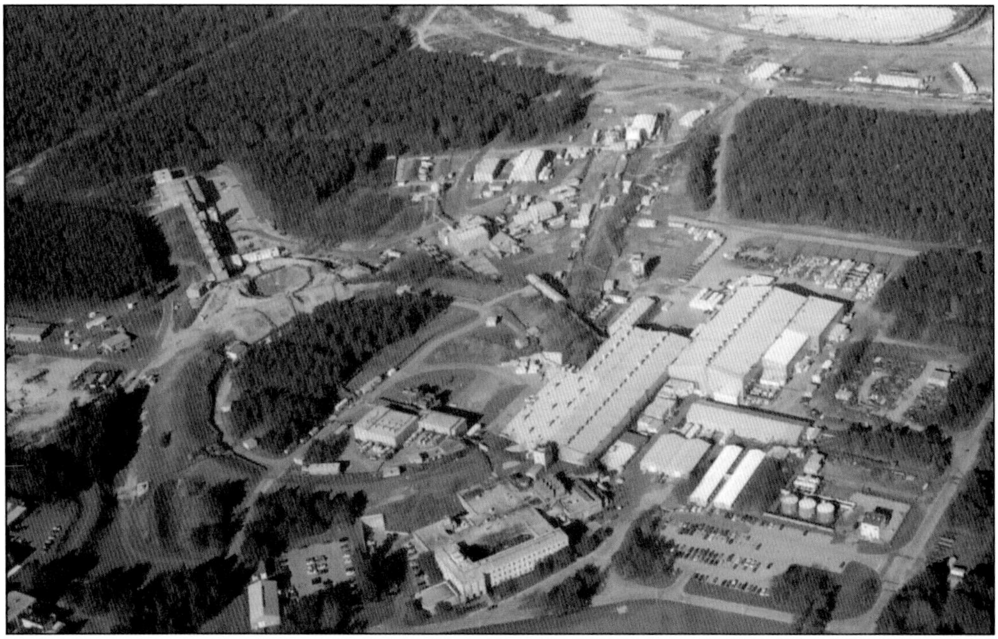

BNL occupies 5,000 of the original 10,000 acres. The remaining 5,000 acres were deeded to New York State for parklands, now Brookhaven State Park. The site is now part of a New York State groundwater protection area, the Central Pine Barrens. In 1981, a Camp Upton Museum was opened in the former Graphic Reactor Building. The museum was unfortunately closed in 2001, and the collection has been in storage since then. Lab employees conduct tours of the site on "Summer Sundays" in July and August.

Bibliography

Adler, Maj. J.O. *History of the 77th Division, Aug. 25, 1917–Nov. 11, 1918*. New York City: W.H. Crawford, 1919.

———. *History of the 306th Infantry*. New York City: The Association, 1935.

Barrett, Mary Ellin. *Irving Berlin: A Daughter's Memoir*. New York City: Simon & Schuster, 1994.

Batchelder, Roger. *Camp Upton*. Boston: Small, Maynard & Co., 1918.

Bayles, Donald, and Paul Infranco. *Camp Upton*. Middle Island, NY: The Longwood Alliance, 2017.

Bayles, Thomas R. *Camp Upton in World War I*. Middle Island, NY: Town of Brookhaven, 1974, revised 1977.

Bergreen, Laurence. *As Thousands Cheer: The Life of Irving Berlin*. New York City: Viking, 1990.

Cary, Roger J. *Camp Upton Convalescent Hospital: A History and Resume of Activities, Sept. 26, 1944 to June 30, 1946*. New York City: S.F. Auerbach, 1946.

Crawford, Gilbert. *The 302nd Engineers: A History*. Foreword by Robert Alexander. N.P., 1919.

Dwyer, Norval. "The Camp Upton Story." *Long Island Forum* (January 1970), 6–10.

Foley, Tricia, and Karen Mouzakes. *Yaphank*. Charleston, SC: Arcadia Publishing, 2012.

The History of the 306th Field Artillery. New York City: Knickerbocker, 1920.

Jablonski, Edward. *Irving Berlin: American Troubadour*. New York City: Henry Holt, 1999.

Miles, L. Wardlaw. *History of the 308th Infantry*. New York City: G.P. Putnam's Sons, 1927.

Ostrow, Dr. Oscar I. "Military Training at Camp Upton during the Great War." *Long Island Historical Journal* Vol 16, No. 1–2 (Fall 2003–Spring 2004), 55–103.

Stallings, Laurence. *The Doughboys: The Story of the AEF, 1917–1918*. New York City: Harper & Row, 1963.

Tiebout, Frank B. *A History of the 305th Infantry*. New York City: The 305th Infantry Auxiliary, 1919.

"Yaphank, Mushroom City of Long Island." *Telephone Review* (October 1917), 290–295.